GOOD PRACTICES IN THE OPERATION AND MAINTENANCE OF LOW ENERGY ELECTROSTATIC ACCELERATORS

The following States are Members of the International Atomic Energy Agency:

AFGHANISTAN
ALBANIA
ALGERIA
ANGOLA
ANTIGUA AND BARBUDA
ARGENTINA
ARMENIA
AUSTRALIA
AUSTRIA
AZERBAIJAN
BAHAMAS
BAHRAIN
BANGLADESH
BARBADOS
BELARUS
BELGIUM
BELIZE
BENIN
BOLIVIA, PLURINATIONAL STATE OF
BOSNIA AND HERZEGOVINA
BOTSWANA
BRAZIL
BRUNEI DARUSSALAM
BULGARIA
BURKINA FASO
BURUNDI
CABO VERDE
CAMBODIA
CAMEROON
CANADA
CENTRAL AFRICAN REPUBLIC
CHAD
CHILE
CHINA
COLOMBIA
COMOROS
CONGO
COOK ISLANDS
COSTA RICA
CÔTE D'IVOIRE
CROATIA
CUBA
CYPRUS
CZECH REPUBLIC
DEMOCRATIC REPUBLIC OF THE CONGO
DENMARK
DJIBOUTI
DOMINICA
DOMINICAN REPUBLIC
ECUADOR
EGYPT
EL SALVADOR
ERITREA
ESTONIA
ESWATINI
ETHIOPIA
FIJI
FINLAND
FRANCE
GABON
GAMBIA, THE
GEORGIA
GERMANY
GHANA
GREECE
GRENADA
GUATEMALA
GUINEA
GUYANA
HAITI
HOLY SEE
HONDURAS
HUNGARY
ICELAND
INDIA
INDONESIA
IRAN, ISLAMIC REPUBLIC OF
IRAQ
IRELAND
ISRAEL
ITALY
JAMAICA
JAPAN
JORDAN
KAZAKHSTAN
KENYA
KOREA, REPUBLIC OF
KUWAIT
KYRGYZSTAN
LAO PEOPLE'S DEMOCRATIC REPUBLIC
LATVIA
LEBANON
LESOTHO
LIBERIA
LIBYA
LIECHTENSTEIN
LITHUANIA
LUXEMBOURG
MADAGASCAR
MALAWI
MALAYSIA
MALI
MALTA
MARSHALL ISLANDS
MAURITANIA
MAURITIUS
MEXICO
MONACO
MONGOLIA
MONTENEGRO
MOROCCO
MOZAMBIQUE
MYANMAR
NAMIBIA
NEPAL
NETHERLANDS, KINGDOM OF THE
NEW ZEALAND
NICARAGUA
NIGER
NIGERIA
NORTH MACEDONIA
NORWAY
OMAN
PAKISTAN
PALAU
PANAMA
PAPUA NEW GUINEA
PARAGUAY
PERU
PHILIPPINES
POLAND
PORTUGAL
QATAR
REPUBLIC OF MOLDOVA
ROMANIA
RUSSIAN FEDERATION
RWANDA
SAINT KITTS AND NEVIS
SAINT LUCIA
SAINT VINCENT AND THE GRENADINES
SAMOA
SAN MARINO
SAUDI ARABIA
SENEGAL
SERBIA
SEYCHELLES
SIERRA LEONE
SINGAPORE
SLOVAKIA
SLOVENIA
SOMALIA
SOUTH AFRICA
SPAIN
SRI LANKA
SUDAN
SWEDEN
SWITZERLAND
SYRIAN ARAB REPUBLIC
TAJIKISTAN
THAILAND
TOGO
TONGA
TRINIDAD AND TOBAGO
TUNISIA
TÜRKİYE
TURKMENISTAN
UGANDA
UKRAINE
UNITED ARAB EMIRATES
UNITED KINGDOM OF GREAT BRITAIN AND NORTHERN IRELAND
UNITED REPUBLIC OF TANZANIA
UNITED STATES OF AMERICA
URUGUAY
UZBEKISTAN
VANUATU
VENEZUELA, BOLIVARIAN REPUBLIC OF
VIET NAM
YEMEN
ZAMBIA
ZIMBABWE

The Agency's Statute was approved on 23 October 1956 by the Conference on the Statute of the IAEA held at United Nations Headquarters, New York; it entered into force on 29 July 1957. The Headquarters of the Agency are situated in Vienna. Its principal objective is "to accelerate and enlarge the contribution of atomic energy to peace, health and prosperity throughout the world".

IAEA-TECDOC-2087

GOOD PRACTICES IN THE OPERATION AND MAINTENANCE OF LOW ENERGY ELECTROSTATIC ACCELERATORS

INTERNATIONAL ATOMIC ENERGY AGENCY
VIENNA, 2025

For further information on this publication, please contact:

Physics Section
International Atomic Energy Agency
Vienna International Centre
PO Box 100
1400 Vienna, Austria
Email: Official.Mail@iaea.org

Printed by the IAEA in Austria
May 2025
https://doi.org/10.61092/iaea.btib-jdpm

IAEA Library Cataloguing in Publication Data

Names: International Atomic Energy Agency.
Title: Good practices in the operation and maintenance of low energy electrostatic accelerators / International Atomic Energy Agency.
Description: Vienna : International Atomic Energy Agency, 2025. | Series: IAEA TECDOC series, ISSN 1011-4289 ; no. 2087 | Includes bibliographical references.
Identifiers: IAEAL 25-01756 | ISBN 978-92-0-106525-4 (paperback : alk. paper) | ISBN 978-92-0-106425-7 (pdf)
Subjects: LCSH: Electrostatic accelerators. | Electrostatic accelerators — Maintenance and repair. | Accelerator-driven systems.

FOREWORD

This publication has been prepared to serve as an aid for the operation and maintenance of low energy electrostatic accelerators and associated infrastructure. It provides information on good practices for the safe and effective operation and maintenance of accelerators, ion sources, beamline apparatus and associated infrastructure and equipment. It also presents practical working knowledge acquired primarily from 'on the job' experiences of many accelerator operations personnel. The material presented has been selected to overcome the general shortage of suitable documentation in accelerator operations, and to present a variety of examples of practical experiences accumulated over many decades of maintaining and operating accelerators.

The scope of this publication is restricted to low energy electrostatic accelerators. These are configured as either single-ended or double-ended tandem types, utilizing chain, belt or cascade charging systems. In the IAEA's Accelerator Knowledge Portal, there are 324 such electrostatic accelerators worldwide, distributed over 59 countries. Given the diverse types of accelerator, each with unique physical and technological characteristics and applications, this publication aims to address general topics of operations and maintenance common to all low energy research accelerator facilities.

An accelerator facility is unique in terms of the large number and the diversity of potential operational and safety related issues at a single location. This includes high voltages, cryogenics, toxic and flammable compressed gases, hazardous materials, ionizing and non-ionizing radiation, heavy equipment and confined working spaces, all of which need to be managed safely and effectively for the protection of personnel and equipment.

This publication provides practical information on good practice and safety considerations, which are prerequisites for the effective management of low energy electrostatic accelerator facilities. It is intended to provide valuable information and knowledge based on best operational and maintenance practices, and on complementing materials from accelerator and related equipment manufacturers, and to serve as an educational and training resource for accelerator personnel and users.

The IAEA would like to thank all the experts who contributed to the drafting and reviewing of this publication. The IAEA officers responsible for this publication were S. Charisopoulos and N. Skukan of the Division of Physical and Chemical Sciences.

CONTENTS

1. INTRODUCTION

1.1 BACKGROUND

The IAEA Member States that are operating low energy electrostatic facilities can have a wide spectrum of locally available technical expertise to draw upon, ranging from laboratories with a large pool of well-experienced and knowledgeable personnel, to new and developing laboratories having very limited personal practical knowledge and experience. In the IAEA's Accelerator Knowledge Portal [1], there are 324 electrostatic accelerators worldwide distributed over 59 countries. Many of these facilities are facing difficulties in training early career scientists and technicians, retraining existing personnel, and enhancing the knowledge and skills of more senior accelerator technicians, engineers, and scientists. The need for education and training on accelerator technologies is evident, and the IAEA is responding to such needs.

Historically, low energy electrostatic accelerators were constructed piece by piece. After the functional requirements were established, major components were sourced from specialized manufacturers, while additional parts were either fabricated in-house or obtained from commercial suppliers. Local personnel then assembled, commissioned, operated, and maintained the accelerator system. The tacit knowledge gained through collaborations between scientists, engineers, and technical staff created a valuable institutional memory. Unfortunately, in many cases, this knowledge was either not documented or inadequately recorded. Documenting workplace activities was often seen as unproductive, with little time allocated for it. As a result, knowledge transfer and dissemination typically occurred locally through mentoring, and nationally or internationally via user forums and symposia. Even today, accessing important information can be challenging, as many people are unaware of what is available and where it is stored.

With the development and maturing of accelerator technologies, components became more reliable, systems became simpler, more versatile, and more standardized, and in the early 1990s, commercial companies began making available 'turnkey' electrostatic accelerator systems, provided and installed with mostly proprietary components and computer control systems. Any upgrades or modifications to these systems are in general provided by the original supplier. Most new and refurbishing low energy electrostatic accelerator laboratories have acquired turnkey systems during the past few decades.

While turnkey systems can fulfil organizational requirements for high productivity and cost-effective operational efficiency, they can diminish the capacity for self-reliance and the sustainability of operations and maintenance by reducing or eliminating opportunities to build and maintain organizational memory.

The IAEA is responding the Member States needs for capacity building in accelerator related nuclear science and technologies through various educational and training courses, e-learning materials, and technical publications. This publication provides a repository of practical working knowledge; knowledge that has been acquired primarily from 'on-the-job' experiences of many accelerator operations personnel. It provides information on good practices for the safe and effective operation and maintenance of accelerators, ion sources, beamline apparatus, and associated plant and equipment. The material presented has been selected to overcome the general shortage of suitable documentation for Member States to support their capacity building programmes in training early career professionals and engineers, and to retrain, upskill, and expand the capabilities of existing personnel.

1.2 OBJECTIVE

The objective of this publication is to provide practical working knowledge and experiences to help overcome the general shortage of suitable documentation for the safe and effective operation and maintenance of accelerators, ion sources, associated plant and equipment, and operational procedures.

1.3 SCOPE

The scope of this publication is confined to topics concerning the operation and maintenance of accelerator systems that use static electric fields to accelerate ions. It does not include applications of accelerated ion beams.

1.4 STRUCTURE

This publication consists of twelve sections. Section 2 describes an effective accelerator facility management system that needs to be implemented to ensure safe and efficient facility operations. Section 3 describes safety awareness considerations and hazards typically found in an accelerator facility.

Sections 4–7 and 11, describe the physical characteristics and the working principles of accelerators and associated equipment, and good practices for their safe and effective operation and maintenance. Sections 8–10 describe the necessary supporting infrastructure and their requirements to enable accelerator systems to function safely and reliably, as well as general issues of their operation and maintenance. Section 12 describes a computer control system for accelerator operators to monitor and control the accelerator system and associated equipment.

2. OPERATIONAL PROCEDURES AND GOOD LABORATORY PRACTICES

An effective accelerator facility management system needs to be implemented to ensure safe and efficient facility operations and to meet all applicable regulatory and legislative requirements. At a minimum, a comprehensive set of Standing Orders needs to be produced that defines responsibilities, authorities, and mandatory work practices. Where a higher level of accountability is required, then a system based on a set of documented Standard Operating Procedures (SOP) would be more appropriate. The main difference between Standing Orders and SOPs lies in their scope and flexibility: Standing Orders are broader directives for recurring or predictable situations, often allowing some discretion within their parameters, whereas SOPs are detailed, step-by-step guides for performing tasks with rigid instructions.

The SOP is a 'living' document that will require ongoing review and change as the accelerator facility evolves. The reviewing interval needs to be at least annually, or when a significant change is made to facility operations. This document needs to be kept in a designated location, and readily available to all staff. Staff need to be encouraged to take ownership of this document and be proactive in further developing it, in response to changing circumstances.

In all cases, the management system needs to be simple, generic, and leave sufficient flexibility to allow for future development without extensive rewriting of documents. It needs to address such issues as responsibilities, safety, normal operation of equipment and maintenance. The management system also needs to track and rectify non-conformances in facility operations, such as operations that fail to meet prescribed standards and equipment malfunctions.

Indicators of an inadequate management system include repetitive failures of equipment, excessive accelerator downtime, high operational costs, inadequate forward planning and resources, and a poor safety record. Such factors can affect the long term viability of the facility. Adopting an existing ISO quality-based management is one method to establishing a credible and effective management system. Where an accelerator facility has an existing management system that works well, and suits the requirements and expectations for that facility, then it might not be prudent to change it.

SOPs are essential for high-energy accelerator installations and, in most cases, are facility-specific (see, e.g. Ref. [2]). For low energy electrostatic accelerator facilities, SOPs are often replaced by Standing Orders. Regardless of whether SOPs or Standing Orders apply to these types of facility, the relevant documentation has to provide clear instructions outlining the steps necessary to implement specific operations. At a minimum, it includes guidelines on:

— General facility operation, including purpose and scope;
— Checklist of tasks before operating the accelerator;
— Accelerator startup procedure;
— Routine operations (checklists and list of authorized personnel allowed in the experimental area when the beam is active);
— Emergency shutdown procedures and emergency protocols;
— Radiation safety;
— Maintenance and inspection;
— Hazardous material handling;
— Reporting and documentation.

To provide general guidelines for preparing the necessary documentation describing the operational procedures at low energy electrostatic accelerators, the following subsections outline some of the elements on which good practices and procedures as based. They are:

- Responsibilities and authorities;
- Safety considerations;
- Normal operation of equipment;
- Maintenance;
- Review and revision.

2.1 RESPONSIBILITIES AND AUTHORITIES

It is essential that an accelerator management system has a clear, and well-defined documented set of responsibilities and authorities. It needs to encompass all levels of the personnel structure, from the most senior to the most junior. Defining the responsibilities at all levels will:

- Facilitate staff to perform their duties more effectively through empowerment;
- Encourage knowledge development;
- Reduce the incidence of demarcation;
- Improve the management of staff resources;
- Enhance the safe management of the facility.

Additionally, authorizing staff to perform specific tasks based on assessed competency will improve areas such as compliance to safety regulations and procedures. It will also provide a mechanism for matching areas requiring specialized expertise to staff competencies. Such a mechanism can be used to promote staff development. In general, there are three main levels of personnel that need to be considered in the overall management structure and assignment of responsibilities: senior manager, facility manager, and operations staff. It is important to ensure the management structure starts from the senior organizational level downwards and includes other groups that have an influence on the daily operations of the accelerator facility. An organizational chart is a useful tool for seeing how the group works, as well as how the accelerator facility is managed.

2.1.1 Senior manager

The person who takes responsibility for the overall accelerator facility needs to be identified. This person is usually at a senior level such as Chief Executive Officer, Director, or Head of Department. They will normally carry the responsibility for ensuring that the group who manages the daily accelerator operations is meeting the expectations, goals, objectives, outcomes, etc. that have been previously set. This person will have as their responsibilities to report the effectiveness and outcomes of the facility to a higher authority, as appropriate.

2.1.2 Facility/accelerator manager

The facility manager has duties that include the daily management of staff and resources, the authority over the budget, responsibilities for ensuring the safe operation of the facility, scheduling of accelerator operations and reporting on their status, and reviewing the management system for improvement.

2.1.3 Accelerator staff

Within this group of people, there may be staff that have specialized areas of responsibility requiring a high level of expertise, knowledge, and experience. Their responsibilities need to be commensurate with their grading, and to utilize their competencies to mentor and guide other staff. It is important that there is a structure within this group that identifies responsibilities to prevent demarcations.

2.2 SAFETY CONSIDERATIONS

Safety is a broad and encompassing area that includes emergency procedures, responsibilities, compliance with operational limits and conditions, external safety regulations, environmental codes of practice, etc. It is very important that a safety regime be adopted that includes training for dealing with potentially hazardous scenarios and conditions. Every accelerator facility needs to have its own internal system of safety management. In some cases, an umbrella system for the organization in which the accelerator facility belongs to, may already include the accelerator facility.

If a safety management system is not in place, the minimum safety issues that need to be planned for by a facility/accelerator manager is effective procedures for dealing with emergencies. This includes fire, exposure to radiation, release of toxic gases and volatile chemicals, and general laboratory and industrial accidents.

With regard to effective safety management, the requirements established in IAEA Safety Standards Series No. GSR Part 2, Leadership and Management for Safety [3] are to be integrated into the facility's management system, wherever applicable.

2.3 NORMAL OPERATION OF EQUIPMENT

The accelerator manufacturer normally provides a basic instruction manual for the operation of their machine; however, this is usually very general in nature and may require supplementary documentation to be developed in-house to ensure compliance with local standards and operating conditions. Such supplementary documentation needs to be written by experienced operational staff in consultation with end users, safety and engineering groups, and regulatory bodies, as appropriate.

Instructions for the normal operation of equipment need to be documented to include all relevant tables, drawings and pictures, that will ensure the unambiguous and safe operation of the facility. These instructions need to be controlled from unauthorized editing and kept in a defined location. Like the SOP, they need to be regularly reviewed for relevance to ensure any changes made are properly recorded. Failure to do this may impact heavily on the safety regime at the facility, and lead to injury to personnel or damage to plant and equipment.

It is very useful to keep a daily log recording operational parameters of the accelerator, including:

- Date and time the accelerator was turned on and off;
- Users' names;
- Terminal voltages;
- Ion source used and ions accelerated;
- Critical vacuum levels;
- Safety related events and accidents;
- Equipment faults;
- Suspected or potential problems.

Logging of operational parameters is useful for other major equipment such as ion sources and gas handling plants. Besides providing information for monitoring equipment performance and maintenance, it provides information that is particularly useful in tracking the effectiveness of the facility management. The information may be kept in a logbook, on printed forms, or in electronic format, but in all cases, it needs to be kept simple. Staff needs to be trained and encouraged to record all significant events, problems, general operating parameters, and comments regarding the performance of the accelerator throughout its operation.

Another useful log to keep is an experimental 'set-up' log that records the experimental parameters required for a particular measurement. This type of log will assist in the reproduction of previous experimental conditions and will reduce the set-up time for similar work. It can be in specific electronic format as in the case of some accelerator computer control system logs, or else in a dedicated logbook. Logged parameters may include:

— Ion source parameters for various ions;
— Charging current versus terminal voltage;
— Typical corona control currents versus terminal voltage;
— Magnetic field or current settings versus ion species and energy;
— Other operational settings for equipment used that may vary from run to run.

2.4 MAINTENANCE

An accelerator's performance is influenced by how well and how regularly it is maintained. Small electrostatic accelerators, like other more complex equipment, require regular preventative maintenance. The interval between maintenance varies depending on the frequency of breakdown faults, prescribed preventative maintenance periods, and economic and budgetary restrictions. Procedures for managing maintenance need to be part of the high level SOP document, while the specific instructions will be contained in more detailed documents that include methods, drawings, tables, pictures, parts lists, references, and troubleshooting techniques.

A key to successful maintenance programmes is the use of logbooks and/or databases to record all maintenance work carried out. It becomes a useful record of all maintenance work previously undertaken and does not rely on the memories of staff to recall the details from, in some cases, many years before. Logbooks need to record the date the work was carried out, what work was done, parts used or replaced, modifications made, and the personnel who carried out the work. This information can be invaluable when trying to diagnose similar problems to those previously encountered.

The mixture of skill sets necessary for reliable servicing depend on the complexity of the accelerator facility in terms of control systems, accelerator type, the nature of the experimental equipment used, and the level of assistance provided by parties outside of the core maintenance group. To ensure continuity in the skills and knowledge needed to operate and maintain small accelerators, it is important that there is more than one person trained to be competent in each of the different maintenance and operating tasks. This will provide coverage of skills for staff that may be absent and provide a strong base for succession planning. Table 1 provides some examples of basic skill sets required for various maintenance tasks.

Consideration of safety is paramount in maintenance, which needs to be planned and conducted only by skilled and authorized personnel. Potential hazards exist in servicing electrical systems, handling hazardous chemicals or toxic materials, and handling equipment that has become activated and/or contaminated. Correct and well-maintained safety equipment is necessary during accelerator maintenance, and needs to include radiation monitors, electrical test meters, lifting equipment, and personal protective equipment. Administrative controls will help prevent unsafe practices and potential accidents in the workplace.

TABLE 1. BASIC SKILL SETS REQUIRED FOR VARIOUS MAINTENANCE TASKS

Maintenance Task	Technical/Engineering Qualifications and Competencies	Knowledge Base
Ion source maintenance	Mechanical, electronic, and electrical instrumentation	An understanding of high voltages, vacuums, handling of volatile materials, ion beam production, radiation
High voltage accelerator	Mechanical, electronic, and electrical instrumentation	An understanding of high voltages, vacuums, radiation
Beam transport systems	Mechanical, electronic, and electrical instrumentation	Vacuums, magnetic fields, high voltages, ion beam production, radiation, ion beam diagnostic and control equipment
Detection systems/end stations	Mechanical and electronic instrumentation	Vacuums, electronic instrumentation, detection systems, compressed gases, mechanical apparatus, high voltages, ion beams, radiation
Gas handling plants	Mechanical instrumentation	Compressed gases, compressors, vacuum, refrigeration
Computer control systems	Computer programming/electronic instrumentation	Control systems, data acquisition systems, radiation
Electrical reticulation systems	Electrician	Instrumentation and control systems, radiation
Site services, e.g. compressed air, water, gas	Specific occupational related qualifications and experience	Vacuums and radiation

It is essential that each facility has responsible safety officers to ensure radiological hazards are safely and effectively managed to minimize the risk of exposure during maintenance. Such hazards include:

- Ion beam reactions, inducing long half-life activations in materials;
- Potentially contaminated maintenance tools and personnel (if not effectively managed);
- Radiation produced from ion beam reactions with certain materials;
- The use of tritium targets to produce neutrons.

If the organization has a radiation protection section that provides personal dosimeters and a recording system, then staff working in designated accelerator 'radiation areas' need to be part of the dose monitoring system. If such a system is not available, then staff needs to wear personal dosimeters and regularly log their exposure. Detailed recommendations on individual monitoring are provided in IAEA Safety Standards Series No. GSG-7, Occupational Radiation Protection [4].

2.5 PERIODIC REVISIONS

Accelerators, once installed and commissioned, may undergo periods of evolution as the user and organizational requirements change. The facilities may also evolve to enhance existing processes, or keep abreast of technological change, so a management system that includes reviewing the facilities for relevance, efficiency, and effectiveness is essential.

2.5.1 Evolution

An effective means of managing the evolving change is by using a system that allows equipment faults and process non-conformances to be identified and tracked. In the ISO 9001 quality management system, equipment and processes undergo regular assessment for their operational condition and adequacy to maintain and comply with a standard process. When a fault or non-conformance occurs, it is documented and acted upon using either a repair or preventative measure, or an action that will correct a repetitive problem. In each case, it is an opportunity to improve or enhance the equipment capability or process.

A simple paper form may be used to record the fault, or non-conformance rectification, and may include at least four sections:

- A section that explains the fault, as recorded by the observer;
- A section that describes the action(s) recommended by the assessor;
- A section that describes the work undertaken to resolve the fault or non-conformance;
- A section that reviews the work done, by both the assessor and the observer to determine whether the work was satisfactory, or whether rework is necessary.

This system is not to be used as a method of generating a list of jobs, as it will fail through overload. It is solely intended to manage faults and processes, and to maintain and improve the overall effectiveness of the accelerator facility.

2.5.2 Planned change

Planned change needs to be implemented as part of good management practices. Sometimes accelerator managers are reluctant to plan and manage on a day-to-day basis, preferring to mentally keep track of progress. This rarely delivers the optimum results. The following points can be used as a guide for establishing best management practices:

- Check that the scope of new work is clear;
- It has a net benefit for users;
- It is budgeted for;
- Resources are available;
- Time is available;
- All personnel who are part of the planning team have provided their inputs;
- A plan is documented and agreed to;
- All credible safety related scenarios have been identified and assessed.

3. SAFETY CONSIDERATIONS IN AN ACCELERATOR LABORATORY

All operational and maintenance activities involve potential safety hazards. Staff undertaking such activities needs to be fully aware of the safety hazards, and the correct procedures to be used to minimize any risk of personnel injury, or damage to equipment. The Standard Operating Procedures (SOP) for all laboratories need to include procedures to deal with accidents, responsibilities for staff in the event of an incident, best practices for the safe operation of the accelerator, abnormal operation of the accelerator, radiation awareness, operational limits and conditions, etc.

The safety culture (see GSR Part 2 [3]) in accelerator facilities can be enhanced by addressing issues such as:

- Emergency procedures;
- Safety assessments;
- Safety reviews and audits;
- Security;
- Designation and classification of work areas;
- Training and retraining;
- Operational limits and conditions;
- Radiation protection and monitoring;
- System and equipment modifications;
- Occupational health and safety.

There are a number of relevant IAEA publications on safety including IAEA Safety Standards Series Nos: GSG-7 [4], SSG-87, Radiation Safety in the Use of Radiation Sources in Research and Education [5]; GSR Part 3, Radiation Protection and Safety of Radiation Sources: International Basic Safety Standards [6]; and IAEA Safety Report Series No. 47, Radiation Protection in the Design of Radiotherapy Facilities [7], that can serve as references also for low energy electrostatic accelerator facilities. In the following, safety considerations specific to these facilities are presented in detail.

3.1 EMERGENCY PROCEDURES

Any situation where the continued operation of the accelerator may be hazardous to personnel and/or equipment needs to be managed with decisive and effective countermeasures. The facility needs to have an emergency shutdown procedure to ensure a safe and controlled shutdown of accelerators and associated plant and equipment. Emergency shutdown of equipment by means of 'kill switches' is of vital importance. An activated shutdown switch needs to render the plant or equipment safe and needs to also indicate that there has been an emergency shutdown. Clearly documented procedures and instructions on what subsequent actions need to follow are to be readily available and known a priori by staff.

Emergency shutdown switches are usually located near the relevant plant and equipment, or in the case of the accelerator, placed around the facility in areas most frequented by personnel for quick action. Examples are near the control or electrical distribution console, and in more remote locations if the facility cannot be entered.

Along with the shutdown switch, there needs to be a documented procedure on what to do in case of an emergency. As an emergency is a rare event, it may be unclear for relatively untrained personnel on the required course of action. The required procedures need to be clear, concise, and unambiguous, and prominently displayed in many frequented locations.

The use of signage with maps, diagrams, pictures, colour, etc. will simplify the emergency procedures, along with a list of numbered prioritized steps written in text with a large font size.

For example:

(1) **PRESS THE EMERGENCY STOP BUTTON**
(2) **CONTACT SAFETY PERSONNEL AND AUTHORITIES** (give names and telephone numbers)
(3) **SWITCH OFF ....**

3.2 SAFETY MANAGEMENT

Developing a safety culture in the accelerator facility can be a long and arduous task, but when entrenched, it becomes the backbone of activities and an important inclusion in planning future activities. It provides confidence that personnel are better able to prevent incidents/accidents and be proactive in addressing potential safety issues. In a well-maintained and operated accelerator facility, there is a good preventative maintenance programme for plant and equipment. Unfortunately, safety risks might not necessarily be managed with the same commitment, usually due to an inadequate or non-existent safety culture. The most comprehensive management system will include a safety risk assessment of the accelerator facility and its activities.

The operation and maintenance of accelerator facilities sometimes requires working under potentially hazardous conditions. Facility managers need to, as part of their management duties, undertake their own risk assessment based on local and international best practices, standards, and ideals. International standards provide a good scope and reference for the expectations and criteria for risk assessments. Some basic philosophies regarding safety risk management include identifying risks, assessment of their potential impacts, how the risk management strategy is controlled, and evaluating the measures put in place to minimize the risk.

The facility manager needs to periodically conduct a safety audit of activities around the accelerator facility to identify any areas where there is a risk of injury to personnel or damage to equipment. An assessment of the activities needs to question:

— What might go wrong?
— Is there anything wrong with the way the activity is being undertaken?
— If the personnel did something differently, how would it change the risk?
— How could the work practice cause injuries or damage?
— What hazards already exist that may contribute to the risk?
— What needs to be done to remove or else minimize the risk?

A risk assessment needs to be done whenever there is potential for a hazardous situation in a work process, or whenever there is a change to normal work practices that could create a hazardous situation. An assessment can and needs to include all staff to capture all relevant information and ideas. It makes the implementation of the risk management strategy acceptable to the accelerator personnel, who also become more responsible for their actions. Managers need to use the assessment as a guide and information for their daily management and control of the facility.

The recommendations arising from the risk assessment need to be evaluated and reported on. Recommendations that are not to be acted on, for whatever reasons, need to be reported on with reasons given. Ultimately, a risk assessment and development of risk mitigation strategies will require time and resources. However, it will help minimize the potential for incidents to occur, and consequently, minimize possible damage to equipment or injuries to personnel.

3.3 SAFETY REVIEWS AND AUDITS

Unlike a risk assessment, safety reviews and audits are ongoing and deal with monitoring the safety compliance of the facility against a set of standards and procedures. It is a way of encouraging and nurturing a healthy safety regime in the accelerator facility. They are usually conducted by a small group of people that come from different work areas, including persons external to the facility. The review team is usually lead by an experienced safety officer. Involving active participation of the accelerator staff in reviews and audits can promote a wider acceptance of the safety culture.

Before undertaking a safety review, the reviewing team needs to ensure there is adequate time and resources available and that the Facility Manager and other key staff are available to be interviewed if necessary. To achieve the best outcome, the facility needs to be divided into small review areas and assessed individually. The outcomes and recommendations of the review, after approval by the Facility Manager, needs to be distributed to everyone who works in the area, and those who have a responsibility in those areas. The review team needs to follow-up their findings and recommendations to ensure that non-conformances are addressed satisfactorily. Reviews and audits need to be carried out at least annually.

3.4 SECURITY

A secured accelerator facility will help protect commercial and research interests, and intellectual property. It may also include controlled personnel access to limit the potential for injuries to staff and visitors unfamiliar with operations. Securing areas may range from warning signs to protective barriers, to controlled access rooms and buildings, and access to computer systems and networks. Often it is an organizational issue that determines the type and nature of security features implemented. The facility manager need only be concerned with local access. Implemented features need to be kept as simple and effective as possible. Staff will invariably reject the use of unnecessary security to restrict access, leading to means to bypass the system.

3.5 DESIGNATIONS AND CLASSIFICATION OF WORK AREAS

Clear boundaries need to be defined to signify potentially hazardous areas. They have to be classified for the level of safety control necessary to maintain a safe working environment. Some hazards such as radiological, electrical, toxic materials, manual handling and workshop practices may have local, national, and international standards that need to be complied with (see, for example, Requirement 24 of GSR Part 3 [6]). By designating a classification to these potentially hazardous work areas, accelerator staff need to apply predetermined safety guidelines when working in these areas.

3.6 TRAINING AND RETRAINING

Safety training needs to include training to deal with abnormal and incident conditions. The level and type of training is dependent on the desired outcomes and performance expectations of the accelerator facility. Training needs to be provided by people who have knowledge and experience commensurate to the subject they are teaching. For an accelerator facility, it is important that the training is divided into areas that involve unique or substantially different types of knowledge. Examples include high voltage generators, ion sources, experimental end stations, data acquisition systems, sample preparation and management, gas/liquid handling plants, etc. Each individual area of training needs to include:

— System knowledge;
— Operational requirements;
— Routine operations;
— Abnormal conditions;
— Practical exercises.

The training syllabus needs to include the responsibilities and authorities for the facility, equipment or process, safety requirements, the different phases of operation, procedures to follow in the case of abnormal operation, alarms, emergency plans, emergency shutdown, etc. The use of training materials such as notes, drawings, diagrams, pictures, manuals, etc. benefits both the staff being trained and the trainer. Tracking the progress of a person undergoing training can be implemented using for example, a simple achievement form that can be progressively dated and signed as the staff member completes each training module. When all modules have been completed, a final assessment may be made to establish their overall competency and success of the training programme. To recognize and acknowledge a staff competency, a 'licence-to-operate' can be awarded. It also provides a system of grading a person's ability to competently operate different parts of the facility.

3.7 OPERATING LIMITS AND CONDITIONS

Setting the upper and lower limits of operation for the accelerator facility and associated equipment will provide a benchmark for operators that will reduce the possibility of damage to equipment and injury to personnel. This may be either a prescribed or engineered limit on the terminal voltage, upcharge current, corona current, ion source output currents, vacuums, radiation levels, pressure of insulating gases, and any other limit that if exceeded could potentially cause damage to equipment or injury to personnel. The facility manager needs to state the operating limits and display these for all to observe and be made aware of.

3.8 RADIATION MONITORING

All accelerator laboratories will produce different levels of radiation depending on their usage and operational conditions and limits. Where it is known or suspected that there are potential radiological hazards in the facility, a radiological survey is required to be undertaken. The radiation survey is required to be conducted in all areas where personnel may frequent during the operation of the accelerator. During the survey, the accelerator operating conditions needs to be maintained to ensure a safe radiological assessment, but sufficient to delineate areas where radiation may be present, and to reliably extrapolate the measured values to those anticipated under normal operating conditions. Recommendations on radiation monitoring are provided in GSG-7 [4].

3.9 MODIFICATIONS

Modifications are changes to existing equipment or plant, to a process, or even to a staff responsibility. Modifications to an accelerator needs to be controlled and recorded to ensure that all users and staff are aware of operational and possibly safety changes, that the modifications are necessary, and that it is done in a way that maximizes the benefits to all interested parties. The method in which the modification is carried out needs to be assessed to minimize the possibility of unnecessary disruptions to the daily facility operations, to prevent the introduction of potentially unsafe work practices, and to complying with regulations and any licensing conditions. All modifications need to be planned and documented.

3.10 OCCUPATIONAL HEALTH AND SAFETY

The operation and maintenance of the facility needs to be done using practices that are consistent with the facility's Occupational Health and Safety (OHS) controls. If the facility does not have the necessary OHS controls, then expert advice needs to be sought.

3.11 SAFETY HAZARDS IN AN ACCELERATOR FACILITY

The following list includes the main safety hazards that may be found in a small accelerator facility:

- Radiation and radioactive material;
- Electrical;
- Mechanical;
- Compressed toxic and flammable gases;
- Noise;
- Manual handling;
- Reactive metals and volatile chemicals;
- Confined working spaces;
- Pressure vessels;
- Radiofrequency and magnetic fields.

These items are now discussed, with recommendations to mitigate any potential hazards.

3.11.1 Radiation and radioactive material

During the production, acceleration and transport of ion beams, X rays, gamma rays, electrons and neutrons can be generated, producing potential radiation exposures to personnel. It is imperative that personnel working in or near the facility have been trained in radiation safety awareness.

The accelerator use will determine the level of radiation protection necessary. All exposure to radiation has to be as low as reasonably achievable and has not to exceed regulatory limits. Personnel exposure to radiation can be controlled by:

- The use of exclusion zones;
- The use of local shielding;

Personnel may become careless if they consider the radiation hazards are negligible. This neglect may lead to improper notice of changed conditions or more serious hazards in certain areas. Careless habits need to be avoided. All radiation producing equipment is required to be respected for their maximum radiation generating capabilities, rather than their normal operating conditions.

3.11.1.1 Maximum permissible occupational doses

The radiation exposure limits need to be set based on local and international standards. The data collected from dose assessments provides a mechanism for managing and controlling occupational exposure of personnel.

3.11.1.2 Personnel exclusion zoning

The person or organization operating the accelerator carries the responsibility for ensuring the safe operation of the accelerator. This includes ensuring radiation levels are optimized, and that access to the operational area is controlled to keep personnel exposure within safe working limits. A simple and effective way for controlled access is by using physical barriers. Two types of barrier may be used depending on the level of radiation present: those that restrict access, and those that prohibit access. A restrictive barrier limits access to areas and are normally used where radiation levels are low but deemed high enough that workers need to avoid not loiter in that area. A prohibitive barrier prevents all access to an area, usually because the radiation levels beyond the barrier are above safe exposure levels. The access barriers need to surround the affected work area but positioned inside of the safe working level boundary. Any local OHS radiation exposure level restrictions need to be considered before choosing the type of barrier.

Local shielding of radiation 'hot spots' will allow greater freedom of movement around the affected areas while limiting any radiation exposures. A radiation survey will identify the areas that may require shielding. The type and thickness of the shielding material is best determined by a qualified OHS officer.

Radiation warning signs need to be placed in clearly visible positions from all directions of approach to warn personnel of the presence of radiation within the barricaded area. The type of radiation, and level and point where the measurement was made needs to be displayed. These signs are to be removed or cancelled by the operator when the radiation hazard is removed. Unnecessary work inside restrictive barriers needs to be avoided.

Where permanent and semi-permanent barriers or other exclusion methods are used, radiation protection interlocks may be installed. The operation of the system needs to be documented, and staff trained in their use. All radiation interlocks need to be regularly tested for correct function and recorded as being such.

3.11.1.3 Monitoring of radiation

All accelerator staff member may be required to wear an appropriate personal dosimeter whenever entering or working in a designated radiation area. Around the accelerator facility, fixed X ray, gamma ray and neutron monitors need to be placed where the main sources of radiation can be produced, to indicate radiation levels to the operator and staff in the area. At a minimum, monitors need to be located near the terminal of the accelerator, around ion sources, and near experiments. Usually, a red flashing or rotating beacon is energized whenever the radiation level exceeds the level set by local standards. The type of units on the monitors needs to be consistent with the local or national standards. It is imperative that radiation monitors are regularly tested and calibrated at least annually. Recommendations are provided in GSG-7 [4].

3.11.1.4 Radioactive sources

Radioactive sources such as gamma or neutron sources may be used to carry out functional tests of radiation monitors around the facility, including experimental instrumentation. All such sources are expected to be encapsulated and clearly identified, and only have a strength sufficient to trip alarm points on the monitors when held beside the detector. They need to be stored in an appropriately shielded and clearly marked enclosure and returned to the enclosure immediately after use. When handling the source, personal exposure needs to be limited by using long handled tongs or shielding, as necessary. A logbook to record source usage is needed to maintain effective control of the sources. Sources need to always be accounted for, and measures need to be taken to prevent them becoming missing or misplaced.

3.11.1.5 Activated materials

Under some ion beam irradiation conditions, beamline or target station components may become activated. This is particularly the case when light ions such as protons, deuterons, or ^{3}He or ^{4}He ions are used. When working on or near potentially activated components, care has to be taken to avoid contamination to the body and equipment (including tools). Activated materials need to be identified clearly and quarantined until they have sufficiently decayed or disposed of.

3.11.1.6 Unsuppressed electron production

High voltage corona discharges can produce X rays, and some types of ion source can produce bremsstrahlung X rays from unsuppressed electron production. In some instances, these sources of radiation are only present during set-up and fine tuning and can go unnoticed. Although the energy of bremsstrahlung X rays may only be tens of keV, it is essential that all areas of concern are adequately shielded to protect personnel from exposure.

Small electrostatic accelerators can produce bremsstrahlung X rays and gamma rays internally due to backstreaming electrons colliding with materials within the accelerator tubes, and from high voltage corona discharges in components comprising the accelerator column and tubes. Some accelerators protect against this radiation by providing an external layer of lead wrapped around the critical areas. Larger machines may use a wall of lead shielding at the area of radiation production or set barriers to restrict access.

3.11.1.7 Ion beams

MeV range ion beams can constitute a serious radiation hazard if not managed effectively. High levels of neutrons can be produced from light ion beams (protons, deuterons, ^{3}He or ^{4}He) colliding with certain elements. It is useful to compile a table listing neutron producing reactions and their yields and keep this as a reference.

Elevated levels of radiation may be produced at locations where the ion beam intersects beamline components such as apertures, slits, and diagnostic elements. Local shielding may be necessary. Deuterons can produce neutrons at low energies and at all locations in the beam transport system, from ion source to end station. They can constitute a particularly hazardous beam. Before utilising deuteron beams, a radiation survey needs to be conducted at low beam currents to ascertain the potential hazards and means to mitigate identified areas of concern.

Exposure to in-air beams is dangerous as the ion beam can deliver extremely high radiation doses to tissue in a very short time. Experiments done with in-air beams need to be adequately safeguarded against accidental exposures. In addition, there can be a potential hazard from ozone production.

3.11.2 Electrical safety

Electrical circuits in accelerator facilities are usually complex and very extensive, carrying a wide range of AC and DC voltages and power levels to different plant and equipment. Only experienced and qualified persons are permitted to work on electrical equipment and supplies. Facility officers are required to ensure personnel in the facility comply with safe electrical practices.

3.11.2.1 Electrical separation

Electrical power circuits have to be separated from signal and control circuits for personnel and instrumentation protection. Breakdown of insulation, surge currents and other abnormal effects in the power circuits can, if not isolated and contained, raise the voltage in the signal and control circuits creating a potential risk of injury to personnel and damage to equipment.

3.11.2.2 Electrical safety

Each country has their own electrical safety standards, but the following is important for any work on supply voltages:

- Only a qualified and licensed electrician can undertake live electrical work on hard-wired circuits carrying supply voltages, that is, any equipment that cannot be isolated by disconnecting the voltage.
- Only a qualified and licensed electrician can isolate hard-wired electrical circuits using circuit breaker switches or fuses at the point of distribution.
- Only a qualified and licensed electrician can remove electrical isolation tags.
- Suitably trained personnel under the supervision of a qualified and authorized electrician can undertake live electrical work on equipment that can be isolated by disconnecting the supply voltage, for the purposes of testing and maintenance.

Good safety practices include:

- Never work alone on potentially hazardous electrical equipment.
- Always ensure circuits are isolated before attempting any work on them.
- Make sure there is adequate light and free access to the equipment, and avoid working in confined or uncomfortable spaces.
- Follow local procedures and national electrical safety standards.

3.11.2.3 Safe working conditions

When working in an accelerator facility there are many obstacles that can hinder access to electrical work areas. They may be platforms, raised plant and equipment, or confined spaces. In all cases, there are access hazards and the potential for a secondary injury from an electric shock by falling, tripping, or bumping is increased. It is important that when working with any voltage that has the potential of delivering an electric shock, the surrounding area is open and clear of obstacles. If this is not possible then consider:

- Moving the equipment to be worked on to a safer area;
- Having another person work with you to warn against potential danger;
- Work on safety insulating rubber mats if working with live wires;
- Refusing to work on the equipment until the area and conditions are made safe.

3.11.2.4 Access to high voltages

Access to high voltage devices is required to be controlled and safeguarded to ensure accidental contact with live or charged components is prevented, such as in ion sources and some beamline focusing and steering elements. Devices such as physical barriers, grounded guards, cages and cabinets, can provide suitable restrictive access, and if necessary, safety interlocked.

Beamline components such as ion sources and pre-accelerators operate in-air with high DC voltages, up to 200 kV. To isolate such devices and protect personnel from contact, a grounded cage which is safety interlocked is needed. The cage needs to surround the equipment at a distance that will not allow an electrical breakdown to reach the cage and be constructed of a wire mesh or solid material of shape and size that makes it impossible for a person's extremities from contacting any high voltage surfaces.

Since most accelerator equipment is operated in open air, an electrical flashover can occur if there is insufficient separation between an exposed high voltage surface and an object at ground potential. As a guideline, all objects at ground potential need to be separated from all exposed high voltage surfaces by a minimum distance of 25.4 mm for every 7.5 kV, e.g. 50 kV requires a spacing of at least 171 mm [8].

3.11.2.5 Warning indicators

All electrical equipment needs to have an indicator light or other visual indication of its operating status and where possible, an isolating switch to isolate the supply voltage. It is important to never assume that because the supply voltage is switched off, the electrical hazard has been removed. Some devices can retain a static electric charge for many hours after the supply voltage has been removed. Additionally, warning indicators might not correctly display the status of faulty equipment.

3.11.2.6 Protective discharging

Before beginning any work on equipment that operates at high voltages (e.g. ion sources, accelerator tubes), or accumulates a charge (e.g. a capacitor bank), the high voltage terminals and components need to be shorted (discharged) to ground to remove any residual charge. A braided ground strap with a heavy-duty metal clip secured to a central ground is a suitable device to discharge objects. The strap needs to be grounded before connecting it to the point to be discharged.

3.11.2.7 Electrical safety practices

The same general precautions have to be taken as with supply voltages however, some special precautions need to be considered. The following represent good electrical safety practices:

— Drawings of electrical circuits and equipment are maintained and kept up to date.
— Before working on any high voltage device, it is shorted to ground. A simple grounding wand with a hook can be used to short out all equipment suspected of holding a static charge.
— Access doors on high voltage equipment or rooms containing charging supplies are interlocked to shut down power and discharge supplies if breached. It is normal to provide more than one interlock, i.e. to provide a backup in case of an interlock failure.
— Power is removed from the high voltage generator and power supplies once the control power key is turned off. The key is removed and kept secure whilst maintenance is undertaken on the high voltage generator.
— The accelerator's high voltage terminal is shorted out with a grounded strap before any work is done.
— High voltage areas are clearly labelled.
— There are multiple warning lights to indicate the "ON" status of high voltage devices.
— Insulators and high voltage cables are routinely cleaned and inspected for damage and replaced if necessary.

3.11.2.8 Isolation

It is necessary to ensure that all circuits can be isolated by a main contactor switch at the point of distribution, and that the circuit has a fuse or a current rated circuit breaking switch. Where possible, residual current detection isolators need to be used at least in areas where electrical maintenance is regularly carried out.

3.11.2.9 Electrical ratings of cable and components

It is necessary to ensure that the cables, plugs, sockets, and fittings are all rated for the conditions that they are to be used. Standard components and cables are not used for purposes other than those for which they are designed. If in doubt, the manufacturer's specifications are to be consulted.

3.11.2.10 Wiring separation

To avoid electrical interference and reduce the possibility of inadvertently short-circuiting power circuits to signal and control circuits, they have to be isolated from each other. In a duct (trench) or cable supporting device (conduit, trunking, trays) the power circuit needs to be physically separated from signal and control circuits by shielding or distancing. If it is necessary for the circuits to share a duct, then the power cables are required to have more than one layer of insulation to reduce the possibility of insulation breakdown.

3.11.2.11 Labelling

All power cables that are run in an accelerator facility need to be labelled and/or coded at least at each end and recorded on an electrical wiring diagram. Permanent signal and control cables also need to be labelled and recorded on a separate drawing. For experimental instrumentation where cabling may be frequently changed, it is useful to produce drawings of the wiring layout and label high voltage cables.

3.11.2.12 Grounding systems

Plant and equipment need to be adequately grounded to ensure correct operation and to protect personnel against contact with energized components. The grounding system has to be comprehensive taking into consideration high and low frequency voltages. All shielding around electrical equipment including the accelerator tank, beamlines, ion sources and end stations need to be linked to the grounding system. Some facilities have more than one grounding system to protect sensitive equipment against intense voltage spikes, as in the case of accelerator tank sparks, and utilize an earth leakage protection system to protect personnel in case of such discharges.

To effectively ground high voltage equipment, a thick multi-stranded copper cable needs to be used and run throughout the facility as the main ground to protect against mains supply and DC voltage discharges, and a large copper braiding or wide flat copper bar with high surface area for grounding high frequency discharges.

3.11.2.13 Electrical insulators

Damage on insulated surfaces may appear as a thin dark line, sometimes not visible, or as a smoke coloured discoloration or burn marks. When checking for a difficult to locate breakdown, it may be best done in a darkened room where the discharging arc or plasma may be observed. It is necessary to always observe from a distance and ensure adequate safeguards against high voltages when undertaking this method.

Insulators need to be regularly cleaned and inspected for damage and replaced if necessary. It is advisable to consult the properties of any cleaning agent before using. Some cleaning solvents if used on insulating materials, such as acrylic plastic, can damage the material, while on others such as Bakelite and mica can trap solvents creating potential fire hazards. If in doubt, it is advisable to consult manufacturers' recommendations.

3.11.3 Mechanical hazards

Mechanical hazards in an accelerator facility are many and varied, and not always apparent. The nature of an accelerator design implies mechanical hazards. The mechanical infrastructure including gas handling plants, air-conditioning, cooling water plants, overhead cranes and lifting equipment, and accelerator components such as ion sources, accelerator pressure vessel, beamlines, and end stations may require frequent intervention to carry out tests, maintenance, research and development. These activities expose personnel to a variety of mechanical hazards. Facility officers have to be aware of the local OHS standards that address many of these hazards.

3.11.3.1 Safe structures

Most accelerators are attached to research organizations that undertake experiments requiring temporary structures. They may be radiation shielding, detection systems, beamlines, platforms, etc. Although they may be temporary, they need to be made in the way to ensure the safety to personnel. A qualified engineer needs to assess the design of the structure before construction begins.

3.11.3.2 Safety risk assessment

A safety risk assessment needs to be made by the facility officer of mechanical hazards in the workplace and addressed by the whole group. This increases staff awareness and promotes a more proactive towards a safe work environment. Some examples of accelerator workplace hazards include:

- Open floor ducts (trenches);
- Unattended ladders;
- Cables, tubes, and ropes on the floor;
- Low and high equipment near passageways;
- Confined spaces;
- Tools and components on the floor;
- Slippery floors, such as oil from vacuum pumps, dripping hydraulics or overhead cranes;
- Vacuum valve actuators that can inadvertently close while being worked on;
- Glass or quartz view ports that can implode shattering and spreading shards;
- Lifting devices and equipment.

3.11.4 Gases and gas handling hazards

Safety hazards with respect to gases found around the accelerator complex may be divided into four general categories:

- Hazards from sulphur hexafluoride (SF_6) with respect to asphyxiation and inhalation of SF_6 breakdown products;
- Hazards of toxic and flammable gases;
- Hazards of handling high pressure gas cylinders;
- Liquid nitrogen hazards.

3.11.4.1 Sulphur hexafluoride

Sulphur hexafluoride (SF_6) is a non-toxic and non-corrosive gas with a density 5 times that of air, has a high dielectric strength, and is a reasonable thermal conductor [9] making it well-suited as an insulating medium in high voltage accelerators. SF_6 is a potent greenhouse gas that has a global warming potential approximately 24 000 times greater than carbon dioxide (over a 100 years) and has an estimated atmospheric residency lifetime of 3200 years [10]. Consequently, many countries have legislation restricting the discharge of fluorinated gases like SF_6. Local authorities need to be consulted prior to any discharges.

When exposed to intense electrical discharges, SF_6 gas decomposes to form sulphur fluoride gases and metal fluorides, which are toxic. If moisture is also present, the decomposition by-products may also include sulphur oxyfluorides, hydrofluoric acid and sulphuric acids [9, 11]. The presence of solid breakdown products can be identified as a white or grey powdery substance and for breakdown gases, a very pungent odour, like rotten eggs.

The safety hazards associated with using SF_6 are inhalation and exposure to toxic breakdown products of SF_6 and asphyxiation due to the displacement of air by SF_6.

To minimize the formation of SF_6 breakdown products, the potential for electrical discharges needs to be minimized, and if they do occur, to reduce the frequency and strength of occurrence. The following actions can help to mitigate electrical breakdowns:

— Monitor tank sparks and reduce the terminal voltage if they persist.
— Never raise the terminal up to high voltages with poor vacuums.
— Check the gas moisture content regularly. Measuring the dew point is a method for determining the gas moisture content. It is important not to operate the accelerator if the dew point is higher than -40°C.
— Slowly condition all high voltage elements in the accelerator after maintenance, or in periods where high voltages are required after low voltages have been used.
— Always condition the accelerator 200–300 kV higher than the required operating voltage.
— Monitor for breakdown discharges along the column, belt or chain, or unbalanced charges on double-ended machines. If problems persist, rectify immediately.
— Periodically check the resistance of column resistors. Replace resistors that have become damaged or have a resistance of more than 10% difference to their original value.
— Periodically check charging voltage diodes. Do not let a situation occur where the first sign of a problem is a low terminal voltage. An uneven voltage gradient may lead to arcing.
— Check the spark gaps on all elements along the column. Clean and adjust if it is apparent that excessive discharging is occurring across the gaps. Ensure that a smooth voltage gradient is maintained, otherwise instabilities may occur.
— Check for breakdown damage along the belt or chain and replace if necessary.

When entry to the pressure vessel is required, precautions need to be taken to minimize potential hazards associated with exposure to residual breakdown products. First, as much of the SF_6 gas needs to be removed from the vessel and transferred back into the storage tanks. To achieve this, a good vacuum of at least 1 kPa needs to be reached before evacuation pumping ceases. The tank is then backfilled with air and one of the entry ports opened. At this time, a measurement needs to be taken of residual SF_6, HF and oxygen. The residual SF_6 needs to be below 1000 ppm and HF below 3 ppm [11] and oxygen levels higher than 19.5% [12]. If the levels are acceptable, then a ventilation fan is used to blow air through the open vessel to ensure the atmosphere inside remains safe to breathe while working inside. If not, then the tank needs to be evacuated again and purged with air until the safety limits are met.

If SF_6 breakdown products are detected in the tank atmosphere, then all the contents of the tank and any open pipework leading to the tank can be assumed to also contain SF_6 breakdown products. In this event, a cleanup of all parts will be necessary, and personnel needs to wear appropriate personal protective equipment. It is possible for SF_6 to escape through seals and accumulate at floor levels where ventilation is low. If enough gas accumulates in these areas, then there is a potential for asphyxiation to occur. The use of an oxygen depletion system is a good method of continuous monitoring for the presence of heavier-than-air gases such as SF_6. Oxygen depletion monitors are best located at floor level, below the accelerator vessel, storage vessel, transfer pipework, or in the gas handling plant.

3.11.5 Toxic and flammable gases

The hazards from the use of toxic and flammable gases are like those found in most chemistry laboratories or industrial plants. Compressed gases that are supplied in cylinders require special attention and handling. The pressure in some cylinders can be as much as 15 MPa and the gases may be toxic or flammable. The cylinders need to be stored in an upright position and restrained to prevent them from falling over. Only safety approved pressure regulators for the gas type and pressure rating need to be used on the cylinders.

3.11.5.1 High pressure cylinders

Fittings for cylinders are designed to be compatible only with components designed for the type of gas in the cylinder. Care has to be taken to ensure all high pressure fittings are assembled as per the manufacturer's recommendations. No attempt needs to be made to continue to use equipment which leaks, malfunctions, or needs maintenance. Such equipment needs to be clearly tagged and set aside for repair by qualified personnel.

Cylinders need to be stored upright out of direct sunlight, in a cool place, and secured against overturning. Cradles are need be used when lifting them with a crane, and cages used to transport them. It is very important that they are not used where they may become exposed to heat sources, spilled cryogenic fluids, molten metal, or an arc welder. The following safety considerations apply when using gas cylinders:

— Cylinders need to be clearly marked, identifying the gas and quality of the contents.
— All gas cylinders are expected to be fitted with a regulating valve so that the outlet pressure can be controlled.
— It is important to never force a sticking valve and keep the valve handle or key on the cylinder valve when it is in use.
— It is necessary to always vent the pressure in delivery hoses when not in use.

Some hazards are peculiar to the use of compressed oxygen gas. The combustibility of materials is greatly increased in an oxygen rich environment. Fittings, lines, and regulators used with oxygen need to be kept free of grease and oil, since only minute quantities in an oxygen rich system can ignite explosively. Only metal gaskets are to be used to seal pipe joints. Oxygen cylinders are required to be stored separately from combustible gases to lessen fire and explosion risks.

3.11.5.2 Liquid nitrogen

The hazards associated with the use of liquid nitrogen are tissue burn from contact with the liquid or vapour, and asphyxiation from evaporated nitrogen displacing the ambient oxygen. Adequate ventilation is required to be available wherever cryogenic liquids are used. The liquids are best handled in dewars or buckets specifically designed for cryogenic use. They are typically a vessel with a vacuum jacket that provides insulation against thermal warming of the liquid. Plastic vessels are not to be used to decant or store the liquids. Liquids are to be decanted below waist height where possible to minimize potential contact from any spillage. It is very important that personal protective equipment is worn when handling cryogenic liquids and, at a minimum, that it consists of eye protection, and cryogenic-rated gloves and footwear.

3.11.6 Hazardous metals, chemicals, and solvents

It is essential that the relevant Material Safety Data Sheet (MSDS) is read, understood, and complied with when handling or using hazardous metals, chemicals or solvents.

3.11.6.1 Hazardous alkali metals

Ion sources can use a variety of hazardous alkali metals for the purposes of sputtering and charge exchange: lithium, sodium, rubidium, and caesium. Alkali metals react violently with water and other chemicals leaving the potential for fire, explosion, and personal injury. When equipment that may be exposed to such metals is cleaned, they need to be dried with the complete removal of water from all contact surfaces. Drying in a vacuum oven is useful but not essential. When handling alkali metals, a closed glove box needs to be used that can be filled with an inert gas such as argon, and reduced in humidity to at least 40%. Some target preparation procedures and vacuum pumping systems utilize hazardous metals such as beryllium, mercury, and cadmium. They are hazardous if inhaled or ingested.

When handling hazardous metals, it is always necessary to use personal protective equipment. Gloves, face shields and eye shields, and if deemed necessary, respirator devices. Operations need to be done in a glove box or in a controlled ventilation system vented outdoors. Inhalation of fumes has to be avoided. All hazardous metals need to be stored in an area where they can be stored safely, and their use controlled. Every material needs to be labelled, or in a labelled container, with unknown metals stored separately until identified.

3.11.6.2 Chemicals and solvents

An inventory of all liquid, gaseous, and solid chemicals, and all solvents used in the accelerator facility needs to be made and kept in a folder along with the MSDS for each item. All chemicals and solvents need to be clearly labelled. All storage areas need to be labelled with safe storage and handling information.

Gloves and eye protection need to be worn when handling chemicals and solvents. Inhalation and ingestion of chemicals have to be avoided. Chemicals and solvents are never to be transferred from their original containers without labelling the new container and ensuring the new storage medium is compatible. Food and drink have to never be stored in cold boxes or refrigerators used for chemical storage. Hands and face need to always be washed immediately after the use of any chemical or solvent.

3.11.7 Noise

Plant and equipment in an accelerator facility can generate noise of varying intensity and frequency. Vacuum pumping systems, belt charge motors, gas plant compressors, workshop tools, cooling pumps, etc. sometimes emit noise at an intensity at which hearing protection equipment is necessary. Whatever protective equipment is worn, always be observant for any warning alarms that may sound while wearing such equipment.

3.11.8 Manual handling

Accelerator buildings typically contain several small lifting capacities and mechanically operated cranes of both the bridge and pedestal type. Cranes and hoists need to be clearly marked as to their rated lifting capacity. The demands imposed by handling heavy or bulky objects in confined areas, and amongst delicate, expensive, and sometimes potentially dangerous equipment, requires that operating personnel be well-trained. A few basic considerations for safe handling operations are:

- Bankspersons and crane operators need to be fluent with the meaning of all directive signals. Some countries may require formal accreditation.
- Cranes need to not be operated unless the crane operator can see the areas to be entered or affected by the load, or can be directed by a bankspersonn who can.
- Only one banksperson has to instruct the crane operator.
- Lifts are not to be made over personnel or over equipment when it can be avoided.
- Tag lines need to be used to keep large objects under rotational control.
- Loads need to not be left suspended on an unattended crane.
- Hard hats need to be worn near crane operations.
- The crane has to not be operated near an accelerator or experimental area without first determining from the area supervisor that it is safe and convenient to do so.
- Lifting equipment needs to be correctly rated for the load to be lifted.
- Lifting equipment needs to not be used if it is damaged.

3.11.9 Confined spaces

Working in confined spaces is a hazard for reasons of lack of normal movement, onset of claustrophobia, bumping, poor ventilation, poor lighting, etc.

It is important that access to confined spaces is controlled, and that persons who need to enter these areas are aware of all the potential hazards. It is also imperative that an emergency plan is in place to rescue a person if they become injured or immobilized. In some countries, it is a requirement to undertake formal training before entering confined spaces, and a permit system is used to control access.

If personnel need to enter a confined space, it is best done with an 'observer' waiting on the outside who can arrange or initiate a rescue if necessary. Also, a sign may be erected to indicate when personnel are inside a confined space and removed once they have exited. The sign may also be used in areas that are not classified as confined space, but in which it is difficult to see or hear when staff are working there.

3.11.10 Pressure vessels

All pressure vessels have to be regularly checked and tested to ensure their integrity is maintained. Large vessels may have to be inspected by local authorities at set intervals and given a certificate of compliance. Some vessels may be required to have a pressure relief valve to vent the contents if the pressure inside rises to a level approaching the maximum pressure rating of the vessel. These need regular inspection and testing.

Personnel can carry out routine inspection of at least the external integrity of the vessels. Things to look for are distortion of the vessel, abnormal leakage in shut-off valves, corrosion on fittings, gauges that appear to be reading abnormal levels or not at all, pressure relief valves that have become obstructed with insects or dirt and dust, and certificates that are out of date.

3.11.11 Radiofrequency fields

Radiofrequency (RF) fields are generated in equipment including some ion sources, power supplies and communication equipment. The RF field is usually restricted to the immediate region using metal shielding, however there are some circumstances where there is little or no shielding. Some single-ended accelerators require the ion sources to be tested prior to sealing the pressure vessel. This requires a plasma to be struck in the ion source while standing near the RF coils. Although it is unlikely that the generated RF field is a safety hazard, care needs to be taken when working near RF devices, especially if pacemakers are worn.

3.11.12 Magnetic fields

Ion beam bending magnets used in an accelerator facility can be large and produce high magnetic field strengths between the pole faces. The magnetic field strength outside the magnet is typically small, however some hazards exist from the mechanical forces of the field on unsecured iron-based metallic objects. Tools and fasteners can become attracted to the magnet and personnel might be hit by flying objects. Persons who wear pacemakers are required not to come near to the magnetic poles. Damage to equipment and personal wristwatches can occur if care is not taken to avoid contact with the magnetic field. It is important that if personnel frequently work near magnets, then signs and/or status lights be erected warning of the presence of magnetic fields.

3.11.13 Seismic events

The general safety considerations in an accelerator facility are commensurate with that found in most industrial or laboratory environments. External factors which may affect safety such as fire and flood are typically covered by the safety umbrella system of the organization that includes the accelerator facility. Resilience of the accelerator facility to seismic events is usually outside the purview of consideration due to the risk of occurrence of significant seismic activity in many geographical locations is extremely low and, the consequences of a significant seismic event appears to be low.

There have been two reported instances of significant seismic events at low energy accelerator laboratories [13, 14]. In both events, significant displacements occurred in the beamlines and beamline elements, however the accelerator itself remained intact and undamaged. The consequences of a significant seismic event on a horizontal, low energy electrostatic accelerator, thus appear to be low.

On 17 January 1995, an earthquake of magnitude 7.2 struck the region of Kobe and Osaka, Japan, with the epicentre a few kilometres from the accelerator facility of the Kobe University of Mercantile Marine, which operates a 1.7 MV tandem-type accelerator. Significant displacements of accelerator system components occurred including beamlines being disconnected and magnets being moved, however the accelerator's internal structure appeared to remain intact and undamaged.

On 22 March 2020, an earthquake of magnitude 5.5 struck the city of Zagreb, Croatia, with the epicentre approximately 10 km from the ion beam facility at the Ruđer Bošković Institute. At the time, the accelerator facility was closed, with all the electronics and vacuum system shut down. During the restart procedure, several vacuum leaks were identified due to beamlines being displaced, and the analyser magnet and accelerator tank were found to have been displaced. After rectification, all the accelerators were able to be restarted and put back into operation successfully, delivering ion beams without difficulty. No damage had occurred to the accelerators.

4. ION SOURCES

Ion sources are devices used to produce charged ions, either positive or negative, depending on the type of accelerator system it is coupled to. In single-ended accelerators, the ion source is located internally within the accelerator vessel, just inside of the positive high voltage terminal. These internal ion sources produce positive ions for acceleration away from the positive high voltage terminal. Double-ended (tandem) accelerators require an external ion source that produces a negative ion beam for injection into the accelerator, and attraction towards the positive high voltage terminal.

This publication discusses only the commonly used types of ion source used for low energy electrostatic accelerators. They are:

- Caesium sputter negative ion source;
- Radiofrequency positive ion source;
- Duoplasmatron ion source;
- Filament driven multicusp ion source;
- Charge exchange negative ion source.

4.1 ION SOURCE SAFETY AWARENESS

All ion sources are potentially hazardous devices. They can contain several high voltage components with exposed surfaces, and can utilize alkali metals such as lithium, caesium, or rubidium, all of which are potentially hazardous if handled incorrectly. Safety awareness of the potential hazards is a prerequisite to undertaking activities on ion sources. The information in this section is intended to help prevent accidents.

All personnel who are required to operate any ion source need to be suitably trained to a sufficient level of competence. Training needs to include knowledge of the recommended and mandatory operating procedures, ion source operations, personnel responsibilities, general laboratory safety, and how to effectively manage abnormal operating conditions.

4.1.1 High voltages

The guiding safety principles for high voltages are fourfold:

- Sufficient protective barriers to prevent persons contacting high voltage surfaces;
- Sufficient safety interlocks and protection systems;
- Adequate electrical insulation to external components;
- Awareness that high voltages may be present through the provision of adequate signage and other visible warnings.

4.1.1.1 Protective barriers

External ion sources are operated in open air, so barriers, cages or other restrictive access devices need to be used to limit personnel access in the immediate vicinity of the ion sources to prevent electric shock. An electrical discharge can travel long distances in air. Adequate electrical shielding is required to be provided along with engineered discharge paths and other protective measures. The dielectric strength of dry air is approximately 3.1 kV/mm at atmospheric pressure at 20°C [15], however this value varies with the shape and size of the electrodes. This minimum distance from high voltage surfaces to ground needs to be greater than the electrical breakdown distance. A normal safe margin is at least double the electrical breakdown distance.

It is important that every external ion source module is enclosed within a protective wire mesh cage, at least 2 m high. This cage height, which is approximately that of a tall person, ought to present a sufficient obstacle to prevent a person from reaching over the top with their hands and touching the ion source, either directly or with some implement in their hand. The cage needs to be a welded and bolted construction using steel tubing for the frames, with mesh welded onto the faces. The mesh size has to be small enough to prevent the insertion of fingers through the wire mesh. The metal cages are fully grounded to the building ground, and wire connections used between all panels of a cage to supplement contacts through bolted connections. In some sections of the wire cage, where better viewing of components and instrumentation is required, Perspex can replace the mesh. A Perspex thickness of 6 mm can offer sufficient electrical insolation.

In general, all components at high voltage need to be sufficiently distanced from the cage perimeter such that a high voltage discharge to a person, or object at ground potential cannot occur. However, where it is necessary for high voltage components to be located close to the perimeter of the cage, i.e. at the point where the beam tube carrying the ion beam leaves the cage, 6 mm thick Perspex may be used to prevent contact with these components.

4.1.1.2 Safety interlocks and protection systems

A shorting rod is used from outside the cage to effectively earth the ion source module before entering. The rod is made from an insulating material with a metal contact probe on the end. The contact probe needs to be connected to ground with a high cross-section flexible copper wire. Before entering and touching any components within a protective cage, all accessible surfaces need to be shorted out using the shorting rod. Simply touching the surface of the components with the shorting rod is sufficient to discharge them, and sometimes a cracking sound can be heard caused by an electrical discharge spark.

Personnel entry to each ion source cage needs to be through a gate, which is closed during ion source operations. Two microswitches need to be mounted on each access gate and wired in series. Two microswitches are necessary to provide sufficient failsafe redundancy in case one microswitch fails. The microswitches have to be interlocked to de-energize all high voltage circuits if the gate is opened or if not properly closed.

Opening any microswitch needs to break the AC circuit to all high voltage power supplies in the ion source module. These switches need to be tested at least monthly, as part of best safety practises. The high voltage supply to each of the ion source modules need to have two mechanisms by which the high voltage reduces to zero after the AC power is removed. First, stored electrical charge will discharge through the power supply's internal circuitry. This could take up to 60 seconds, depending on the discharge time constant of the equipment. A second discharge path needs to be provided through the resistors used across insulators that hold the ion source voltage above ground potential. These discharge paths are in parallel.

Power supplies need to be checked to ensure they are current limited, so if personal contact accidentally occur, the maximum current flowing would be limited and not be injurious. On a well-designed ion source platform, with current limited power supplies, the typical operating leakage currents from high voltage power supplies may be 1–10 mA.

It is essential that key lock switches on the AC supply to the ion source module are installed. This lockout safety feature will prevent persons other than the keyholder from inadvertently energizing the ion source module.

4.1.1.3 Insulation

Ion source components typically need to be manually adjusted and controlled from a location external to the caged area. Three commonly used methods of implementing this are:

- Insulating rods: These are used to extend the ion source power supply control knobs to the outside of the cage. They are usually made from either Perspex or polyethylene and have a sufficient length to exceed the voltage creep distance. Fifty centimetres is typical the minimum length used, but this value this is dependent on the voltage being insulated against. Where the rods have a metallic knob fitted outside the cage, the rod is passed through an earthed metal ring attached to the wall of the wire cage. In the event that the rod becomes conductive, there is an effective earthed protection point between the high voltage and the person. It is worth noting that electrostatic equipment very quickly attracts dust and in the case of carbon particles in polluted air, the insulating rod can become more conductive than anticipated. It is therefore of utmost importance to always keep the insulating rods clean and remove any dust from any surfaces that can come into contact with the personnel during normal operation.
- Fibre optic cables: Sheathed fibre optic cable can be used to transfer digital signals between ion source modules and the external control instrumentation, which is at ground potential. When selecting a suitable optical fibre, its dielectric strength needs to be at least 2 kV/mm.
- Plastic tubing: Plastic tubing such as Nylon, polyethylene, and Teflon is used to carry compressed gases and liquids to the ion source module. The dielectric strengths of plastic tubing have to be of high dielectric strength, at least 2 kV/mm, to provide sufficient electrical insulation. The dielectric strength of the liquids or gases being carried inside the plastic tubing need to also have a high dielectric strength.

4.1.1.4 Signage and warnings

All cages need to have attached signage, using logos and words to indicate high voltage hazards within. Further signage in the general area of the cages need to warn 'High Voltage Area' to exclude personnel other than accelerator staff.

A flashing strobe light needs to be visible from all viewing angles, and from a distance to alert personnel that power is applied to the ion source module and that high voltages may be present. Conspicuous warning lights mounted on the perimeter of the caged area need to be connected to automatically turn on whenever power is applied to the ion source module. A further set of warning lights can be mounted on the ion source module platform and powered from the secondary side of the isolation transformer.

4.1.2 Safe handling of alkali metals

Ion sources can use a variety of potentially hazardous metals, such as lithium, sodium, caesium, or rubidium, for the purposes of sputtering and charge exchange. It is imperative that the relevant MSDS is read, understood, and complied with when handling or using such metals. These alkali metals decompose violently on contact with water and other chemicals, leaving the potential for fire, explosion, and injury. When handling, a closed glove box needs to be used that can be filled with a dry inert gas such as argon and reduced in humidity to at least 40%.

It is essential to use protective clothing when handling these ion source metals. They react readily with moisture or skin to form a metal hydroxide that can cause chemical burns. If the metal ignites, it can cause thermal burns. If it is necessary to heat these metals, it needs to be done in a glove box, or in a controlled ventilation system vented outdoors. Inhalation of fumes need to be avoided. All hazardous metals need to be stored in a cool, dry area where they can be safely stored, and controlled to stop accidental misuse. Every ampoule of alkali metal needs to be labelled as to its contents or stored in a similarly labelled container.

4.2 CAESIUM SPUTTER NEGATIVE ION SOURCE

A caesium sputter ion source creates negatively charged ions from the sputtering action of a caesium ion beam on a solid material. The material to be sputtered is mounted on a negatively biased cathode, usually made from copper or aluminium. Negative ion beams of most elements can be produced; however, the ion beam currents are highest for those elements with a high electron affinity. A well-maintained and well-operated ion source can produce high quality ion beams for hundreds of hours, enabling the accelerator facility to be effectively utilized for long periods of uninterrupted operations.

4.2.1 Principle of operation

The principle of operation of a caesium sputter ion source will be described with reference to Fig. 1.

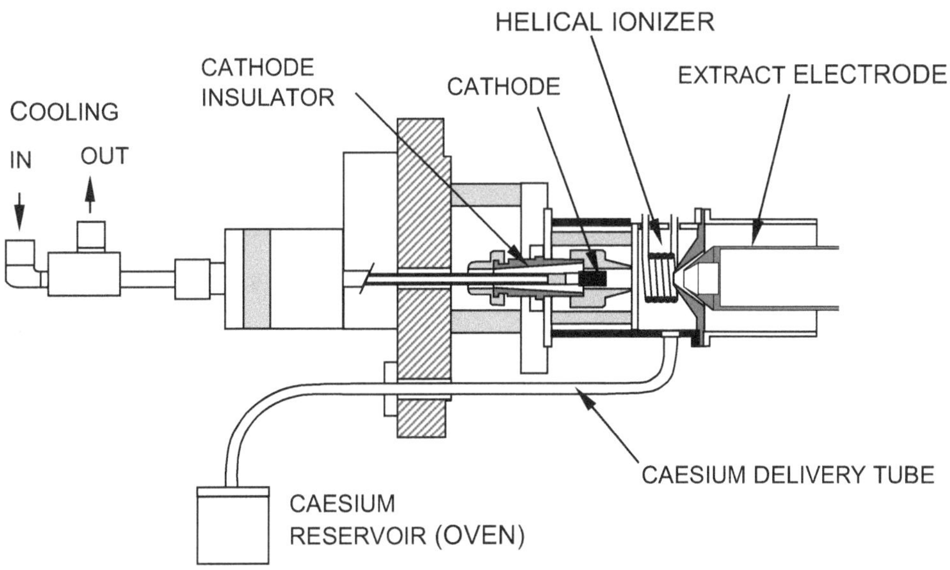

FIG. 1. A schematic of a single sample, caesium sputter negative ion source.

Caesium metal is contained in the reservoir below the ion source body which is heated to produce a flow of caesium vapour. This caesium vapour travels along a delivery tube to an enclosed volume between the cooled cathode and the heated helical ionizing surface. A recirculating coolant keeps the cathode cool. The caesium vapour is thermally ionized on contact with the hot ionizer surface producing positive caesium ions. The cathode, which is negatively biased, attracts the positive caesium ions and upon impact, sputtered ions of the cathode material are produced. Some materials will preferentially sputter negative ions, while others will preferentially sputter neutral or positive particles. The latter can attach electrons as they pass through the condensed caesium layer on the cathode's surface producing negative ions. The negative ions produced are attracted towards the positive extract electrode and after exiting, are subsequently focused, and accelerated by a pre-accelerator. A turbomolecular vacuum pumping system maintains a vacuum pressure of typically 0.1 mPa inside the ion source.

4.2.2 Preventative and corrective maintenance

Preventative maintenance is an effective means of ensuring continued and reliable operations. Table 2 lists the main common caesium sputter ion source components that need to be monitored and routinely inspected. Details of all maintenance work done needs to be recorded in a logbook.

TABLE 2. PREVENTATIVE MAINTENANCE OF CAESIUM SPUTTER ION SOURCES

Item	Maintenance
Caesium reservoir level	Change or replenish caesium when it is noticed that the output current is low or has an erratic behaviour.
Cathode coolant level	Keep coolant above the minimum operating level. Regularly inspect the coolant recirculating pump for sufficient flow rate.
High voltage cables	Regularly inspect for electrical breakdown damage and replace if necessary.
Extract insulator	Keep surfaces clean, inspect insulating surfaces for breakdown damage.
Operating platforms	Keep platforms clean to minimize extraneous electrical leakage paths.
Vacuum seals	Regularly inspect seals around the cathode tube for wear and replace if necessary.
Power supplies	Monitor voltages and currents in all power supplies for abnormal changes. Investigate unexpected deviations from normal operating conditions.
Caesium oven	Monitor oven temperature and investigated unexpected deviations.
Sample carousel	In carousel type multiple sample ion sources, inspect the spring clip arrangements on the carousels that lock the cathodes in place. A loose clip may lead to catastrophic failure of the ion source during cathode loading. Do not use defective cathode sections.
	Inspect the split fingers used to pick up the cathodes and insert them into the ion source. The fingers can bend with use. Damaged fingers have to be replaced.
Internals	Clean ion source internal components, including the ionizer, cathode hat, and insulator after prolonged ion source usage. Scour surfaces using fine abrasive paper to polish, or grit blast. Do not use chemical polishing agents. Use ethanol and or acetone solvents to finish, and dry.

4.2.3 Faults, possible causes and rectification

There are many and varied faults which can occur in ion sources. Table 3 lists commonly occurring faults and suggestions for their rectification. The information is a guide only and can supplement the manufacturer's recommendations.

TABLE 3. LIST OF POSSIBLE FAULTS AND RECTIFICATION SUGGESTIONS FOR A CAESIUM SPUTTER ION SOURCE

Fault	Possible Causes	Rectification tips
No beam output	No beam measurement	Faulty Faraday cup or Faraday cup control system. Check Faraday cup wiring, coaxial connections, signal continuity. Faulty microammeter or input cabling. Put a test current through the meter to confirm output. Check cable and connections. Recalibrate meter.
	Low caesium flow	Blockage in the delivery system, oxidation of caesium due to air ingress, caesium fully consumed, or the oven is not working. Incorrect operating temperature. Check for vacuum leaks. Remove and completely clean the oven and delivery tube. Replenish caesium.
	Loss of ion source high voltage power supplies required for extraction of the ions	Internal power supply fault, faulty high voltage connection or excessive power supply loading from ion source electrodes shorting, or poor vacuum. Individually raise the voltage on the power supplies and monitor the current. If excessive current is drawn, then disconnect the power supplies and check the input from the ion source for shorting or low resistance. Check vacuum system operation. Check ion source for vacuum leaks.
	Loss of caesium ionization	Ionizer, or ionizer power supply failure. Remove and clean the ionizer. Grit blast for the best results. Be careful not to damage wiring during cleaning. Check power supply as for the power supply checks above.
Unstable beam output	High voltage not being maintained	Check power supplies and insulators for breakdown.
	Contaminated internal surfaces	Excessive caesium condensation requires cleaning. Cleaning needs to be scheduled after a set usage time to prevent potential problems.
Multiple sample loading (carousel type holders)	Alignment of the optical reference point has moved	Realign and retest before placing back into service.
	Carousel is contacting the vacuum box interrupting the rotation of the encoder	Check shaft alignment and carousel 'roundness'.
	Spring clips in the carousel not holding the returned cathode in place	Mark the damaged holes and do not use.
	Fingers on the cathode pick up shaft are bent causing unreliable 'pick ups' or 'drop offs'	Replace fingers.
	Slipping drive gear on toothed belt causing encoder position signal to be interrupted	Tighten toothed gear or replace if damaged.
X ray production	Excessive production of X rays	Electrons are not effectively suppressed in the ion source. This will occur if the suppression magnets are removed or incorrectly placed from their normal location.

TABLE 3. LIST OF POSSIBLE FAULTS AND RECTIFICATION SUGGESTIONS FOR A CAESIUM SPUTTER ION SOURCE (continued)

Fault	Possible Causes	Rectification tips
Low beam output	Inaccurate beam measurement	Faulty Faraday cup. Check all connections. Check the alignment of the cup relative to the incident beam. Ensure the leads are connected to the expected sections of the cup. Check beam stability. Faulty microammeter. Check the microammeter as mentioned above.
	Low caesium flow	Low caesium supply or low oven temperature, true or indicated. Turn off the caesium heater and check for changes in the output. Little or no change will indicate a blocked delivery tube or exhausted supply of caesium. Open oven, clean and replenish with caesium.
	Poor ionization of caesium	Low ionizer temperature or contamination of the ionizer surface. Check cleanliness of the ionizer. Check for damage to the ionizer leads. Check power supply output.
	Neutralization of negative ions	Poor vacuum in the ion source. Compare operating vacuum to base vacuum. Isolate the ion source by shutting valves. Leak test and make repairs as necessary. Clean inside of ion source.
	Loss of focus of the ion beam	Breakdown of extraction voltages, breakdown of Einzel lens voltages or breakdown of pre-acceleration voltages, all of which can be due to respective electrode loading or shorting. Power supply loading can occur if the cooling medium has become conductive due to contamination. If the voltage gradient resistors change value or become open circuit on the pre-acceleration tube, poor focusing may result.
	Loss of focusing power of the Einzel lens	Broken grid (if used).
	Insufficient cooling on the cathode	Check coolant flow.
	Loss of sample material in cathode	Can be checked by analysing the ion beam to ensure ions are not being generated from the cathode.
	Poor alignment of cathode in X, Y or Z axes	X/Y alignment is usually only applicable to multi-sample sources. The Z-axis alignment is crucial as an ill-aligned cathode can sputter the sides of the cathode if too far in, or too much of the front cathode surface instead of the centre sample material.

4.2.4 Changing the caesium in the ion source

4.2.4.1 Removal of the caesium reservoir

The following steps are suggested for removal of the caesium reservoir:

- Isolate the ion source from the beamline and the vacuum pumping system by closing the isolating gate valves.
- Admit dry argon gas into the ion source, being careful not to also admit air. A good practice is to first evacuate the hose from the bottle of argon gas to the ion source gas admittance valve with a roughing pump before starting the flow of argon gas.

- When the ion source is fully vented to atmospheric pressure, reduce the flow using the gas regulator to a slow bleed to keep a slight positive pressure inside the ion source to stop air from entering.
- Ensure that the cap for the caesium delivery tube is clean and dry.
- Whilst holding the caesium reservoir, undo the screws from the source body.
- Slowly withdraw the caesium reservoir until the end of the delivery tube is at the opening in the source.
- Place the cap over the end of the delivery tube.
- Place the reservoir in a safe storage place, ensuring that it remains upright.
- Plug the delivery tube hole in the ion source, stop the argon flow, and close all gas admittance valves.

If the capped oven is to be removed from the source for more than a few hours, the assembly needs to be stored under an argon atmosphere. This is to ensure that the caesium does not oxidise due to leakage/permeation of air into the reservoir.

4.2.4.2 Removal of the unused caesium in the reservoir

The remaining caesium in the reservoir needs to be decanted into a container, inside a glove box which has an inert atmosphere. To remove any remaining droplets of caesium that may be present, the caesium reservoir is removed from the glove box and, with caution, placed into a large container of water. Any remnants of caesium will spontaneously combust on contact with the water.

4.2.4.3 Cleaning the reservoir

Cleaning the reservoir involves the following steps:

- Clean the reservoir in water with a scour pad, or fine grade abrasive paper.
- Dry it with lint free wipes, rinse in acetone, and wipe with lint free wipes to remove any residues.
- Dry with a hot air supply. Avoid touching the cleaned inside surfaces with bare hands.
- Flush the delivery tube with the aid of a squeeze bottle containing a mixture of 30% nitric acid and 70% water, then flush with clean water, preferably demineralized, and dry using hot air.
- Store the reservoir and delivery tube in a clean dry plastic bag until needed.

4.2.4.4 Loading new caesium into the reservoir

As previously mentioned, caesium is a potentially hazardous material if handled incorrectly. The loading of reservoir needs to occur in a glove box suitable for the required operations, continually purged with a dry inert atmosphere, and with persons wearing suitable safety wear including protective gloves and eye protection.

For optimum results, it is important to minimize the time the caesium is in contact with the ambient inert atmosphere. The filling procedure needs to proceed without undue delays; thereafter, the filled caesium reservoir be immediately re-attached to the ion source. This requires that an efficient procedure be established in which all equipment and tools to be used are immediately available when needed.

The following procedure is for a caesium sputter source where a reservoir has a flange with delivery tube connected. The basic methodology is to heat the caesium in the ampoule until it flows freely (melting point is 28.5°C), break the neck of the ampoule and pour its contents into the reservoir. The procedure is repeated until the reservoir is full. The delivery tube is screwed into the reservoir, capped with the plug, then transported to the ion source and attached.

- First prepare the glove box. The required charge of caesium ampoules is placed inside the glove box, along with a suitable container for depositing the emptied ampoules, and a small sharp-edged knife to score the neck of the glass ampoules. The cleaned and dried caesium reservoir is also placed inside the glove box, preferably held upright by a laboratory clamp. The reservoir delivery tube and a plug to seal its end are also placed in glove box. Close the glove box and purge it with a dry inert atmosphere such as argon.
- Heating of the caesium is easy achieved using a 'hot block', which is a heated solid metal cylinder with a central hole large enough to accommodate one ampoule. A typical cylinder made be made of brass, ~100 mm diameter and ~25 mm in length. The 'hot block' is heated externally to the glove box and when hot ~50°C, transferred into the glove box. Score around the neck of an ampoule, then place inside the hole in the heated block.
- When the caesium metal has noticeably melted (i.e. it runs freely inside of the ampoule), snap off the top of the ampoule and immediately pour the contents into the reservoir. Be careful not to splash any caesium onto the vacuum sealing surface on the reservoir. Sometimes the ampoule will need a careful downward shake to release the metal. Once it starts to pour it will flow freely. Place the emptied ampoule in the disposal container and repeat the filling process for additional ampoules. The 'hot block' needs to have sufficient heat capacity to melt all ampoules.
- When the reservoir is filled, and with a new clean seal, attach the delivery tube flange, insert the screws, and tighten, and insert the delivery tube plug. The reservoir is now ready to transport to the ion source.

4.2.4.5 Installing the reservoir onto the ion source

The ion source needs to be vacuum leaktight and already purged with argon before starting to load the reservoir. Then, the procedure is as follows:

- Start the ion source's vacuum pumping system to have it on standby, ready for immediate use after attaching the caesium reservoir.
- With argon flowing into the ion source chamber, remove the seal from the reservoir port and wait a few minutes to ensure a complete system flush has occurred.
- Remove the capped reservoir from the glove box and transport to the ion source. Hold the outlet of the delivery tube close to the reservoir port on the ion source, remove the cap and fit the delivery tube inside the ion source.
- Place the screws in position and tighten to make a vacuum seal. Halt the argon flow and evacuate the ion source with the vacuum pumping system.

4.2.4.6 Conditioning the caesium

During the filling of the reservoir, caesium may solidify on its cool sides trapping air bubbles. These trapped bubbles, and gases dissolved in the caesium, need to be removed before raising the reservoir to its normal operating temperature. Warming the reservoir to ~50°C for a time before beginning normal operations can achieve suitable degassing.

4.2.4.7 Disposing of the residual caesium in the used ampoules

The procedure for disposing of the residual caesium in the used ampoules is as follows:

- Place the used ampoules on their side in a slightly elevated position, in a dry sink in which there is a plug in the sink drain.

— Connect a plastic tube to the water outlet with the other end lying freely at the bottom of the sink.
— Turn on the water flow to produce a trickle feed. The water level will slowly rise and eventually enter the ampoules. A few small cracks and pops are usually heard as the residual caesium combusts as it contacts the water.
— The ampoules may be safely removed after they are fully flooded with water.

4.3 RADIOFREQUENCY POSITIVE ION SOURCE

Radiofrequency ion sources produce positive ion beams for use in single-ended accelerators. Protons are produced by a RF field ionizing hydrogen gas, and singly charged helium ions from ionizing helium gas. Other gaseous ions can be produced using the appropriate feed gas.

4.3.1 Principle of operation

Figure 2 shows a schematic of a RF positive ion source. The ion source is constructed from a glass bottle, usually Pyrex or quartz, with a small aperture at the base end, termed the canal, through which the ion beam passes. At the other end of the glass bottle is an anode, termed the probe, which is either a flat plate or thick straight wire. The probe is charged positively, attracting electrons and assists in extraction of the positively charged ion beam from the ionized plasma.

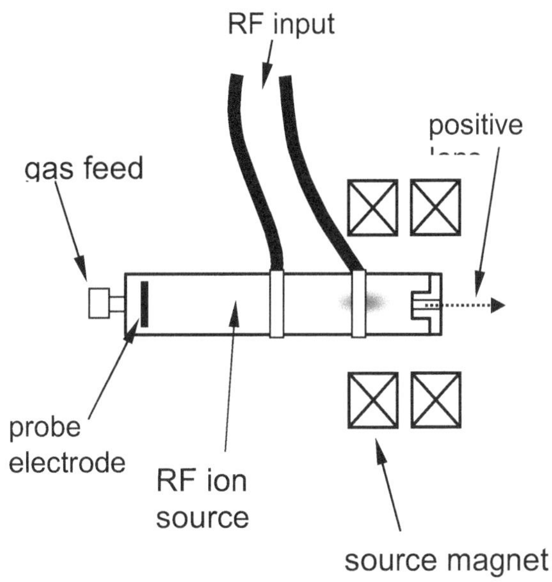

FIG. 2. Schematic of a RF positive ion source with capacitive coupling.

The gas flow into the ion source bottle is regulated to establish a constant pressure of a few mPa. An RF field of approximately 100 MHz is capacitively coupled via two circumferential electrodes mounted toward the canal end of the source bottle. The alternating RF field excites free electrons into oscillation ionizing the gas producing further electrons until eventually, a gas discharge is produced. Electrons are attracted to the probe end of the ion source bottle by a positive bias, whilst the positive ions are attracted towards the canal end of the ion source bottle.

An electromagnet mounted at the canal end assists in the formation of the plasma in the bottle, minimizing oscillations in the plasma and assisting in extraction of the ions through the canal. Some adjustments of the magnet location may be required to optimize or tune the ion source for maximum output. Materials used in the construction of the ion source are selected to provide low plasma recombination and sputtering rates. When generating hydrogen ions, an aluminium canal is used and when helium ions and heavier ions are generated, the canal material is changed to tantalum.

The plasma created inside the ion source bottle will vary in colour depending on the gas used. Hydrogen will produce a reddish-pink colour and helium a pale blue colour. Immediately after maintenance of the ion source bottle, it is common to initially see a greyish-blue colour which is produced by remnant nitrogen that has not yet been flushed from the ion source bottle.

Radiofrequency ion sources typically operate for several hundred hours when used to generate proton beams at 30–50 μA. After this time, the bottle will be seen to have darkened considerably, and if an aluminium canal has been used, the canal will appear eroded at its inner end. The eroded canal material contaminates the inner surface of the bottle causing increased recombination in the plasma, thus reducing the ion source output. It is possible to refurbish ion source bottles using specialized cleaning techniques discussed later in this section.

4.3.2 Faults and possible causes

Table 4 lists common faults and possible causes in an RF ion source. The information is a guide only and can supplement manufacturer's recommendations to rectify faults found in RF ion sources.

TABLE 4. FAULTS RELATED TO A TYPICAL RF ION SOURCE AND THEIR POSSIBLE CAUSES

Fault	Possible Causes
No beam output	No beam measurement due to faulty Faraday cup, faulty microammeter or input cable.
	Loss of gas feed due to fault in metering valve (variable leak) or depletion of the feed gas.
	Loss of extraction or focusing high voltage.
	Loss of the RF power due to oscillator failure or output triode failure.
Low beam output	Inaccurate beam measurement due to faulty Faraday cup or faulty microammeter.
	Dirty or contaminated Pyrex glass bottle.
	Extraction canal eroded or partly blocked.
	Low output from RF triodes.
	Insufficient gas feed.
	Loss of focus of the ion beam due to breakdown of extraction voltages, breakdown of Einzel lens voltages, or breakdown of pre-acceleration voltages.
	Balance between the probe and extraction voltage is wrong.
Unstable beam output	As for low beam output but with a fault of an intermittent nature.

4.3.3 Maintenance of gas discharge bottle and canal

The ratio of positive ions formed in the gas discharge bottle to the amount that is extracted is variable. There are many parameters that can change this ratio including gas pressure, contamination in the bottle from sputtered canal material and hydrocarbons from vacuum oils, and probe voltage fluctuations due to the contamination in the bottle causing resistive changes between the probe and ground.

Ionized atoms are inherently in an unstable state. When materials are exposed to the plasma, they act as catalysts for the recombination of ions, neutralizing them. At the low pressures that RF ion sources operate at, the mean free path of the ions are typically larger than the bottle's dimensions, so surface recombination effects predominate over volume recombination effects. To minimize the surface recombination, the bottle needs to be made of a material that has a low recombination coefficient. The recombination coefficient is highest for metals (~1) and lowest for silicate glasses such as quartz and Pyrex (~10^{-4}).

To decrease the surface recombination rate of ions at the exit canal, a quartz sleeve is used to act as a virtual anode and focus the positive ions into the channel. Over time, canals will become eroded due to sputtering caused by the impact of ions on its inner surface and end. This sputtered material is deposited onto the inside of the quartz sleeve causing a gradual increase in the fraction of molecular ions in the output ion beam and concomitantly, a gradual reduction in the useable ion beam current. The molecular fraction can increase to many tens of percent after hundreds of hours of operation. Eventually, the system needs to be disassembled, cleaned, and if necessary, the canal needs to be replaced if there is excessive erosion.

4.3.3.1 Cleaning the Pyrex discharge bottle

Cleaning the Pyrex discharge bottle involves dissolving the surface deposits with a solution of concentrated nitric and hydrofluoric (HF) acids, as follows:

Warning: This procedure requires working with hazardous chemicals that may cause fatal injury if mishandled. In the event of a hydrofluoric acid burn occurring, seek emergency medical help **immediately**. All procedures need to be undertaken in an acid rated glove box, with full personal protective equipment including eye protection, two layers of acid rated gloves, and acid proof apron. Do not undertake this procedure alone. Ensure safety personnel are aware that you are about to use HF acid, and ensure you understand the dangers and the safe handling techniques as stated in a Material Safety Data Sheet for HF acid.

— First, remove the ion source bottle from the aluminium base by heating in a small oven to ~200°C. This will soften the glue to aid release.
— Next, prepare a solution consisting of by volume, 80% HF (40%) and 20% HNO_3 (100%) in a HF rated plastic beaker, and fill to a depth enough to cover at least ⅔ the length of the bottle.
— Place the bottle into the solution and gently agitate without splashing. The deposited layer will dissolve within 10 to 15 minutes.
— Remove the bottle, rinse in clean water, and then place the other end in the solution until it is also cleaned. Remove and wash thoroughly in clean water.
— After cleaning, the source bottle needs to be washed thoroughly again with distilled water and baked at 300°C for 30 minutes to dispel any remnant water and to relieve any stresses in the glass. Any remaining brownish tint in the glass is most likely due to radiation damage and may disappear when the glass is annealed. The same procedure is required to clean the quartz sleeve.

Caution: Metal components including the extractor need to not be etched with the HF + HNO_3 solution. They can be washed in water with an abrasive scour pad or fine abrasive paper. Rinse off with clean water, ethanol and then acetone. Wipe off excess solvent and dry with a lint free cloth and hot air. Store parts in a clean plastic bag prior to use. Other components can be washed in a similar way, except care needs to be taken not to scuff insulator surfaces.

4.4 DUOPLASMATRON ION SOURCE

The duoplasmatron is an ion source capable of producing high intensity beams of light ions from gases such as hydrogen and helium, and with a high brightness.

4.4.1 Principle of operation

The principle of operation of a duoplasmatron ion source will be described with reference to Fig. 3. A cathodic plasma discharge is created by a few tens of volts between a thermionic cathode and an intermediate electrode acting as a primary anode. The gas pressure inside this section of the duoplasmatron is relatively high, ~100 Pa. The cathodic plasma is guided by a strong axial magnetic field through an aperture within the intermediate electrode into the second discharge chamber, hence the name, duoplasmatron. In the second chamber, the plasma discharge is created between the intermediate electrode, now acting as a cathode and the main anode. Here, the gas pressure is much lower (~1 Pa), and the discharge voltage is higher (~100 V). The anodic plasma created in this second stage is accelerated by the electric field between the intermediate electrode and the anode and is extracted through a small aperture in the anode into a third chamber, the so-called expansion chamber, as the ion beam is unfocused. A duoplasmatron yields atomic, molecular, and triatomic ions, which are discriminated by magnetic analysis.

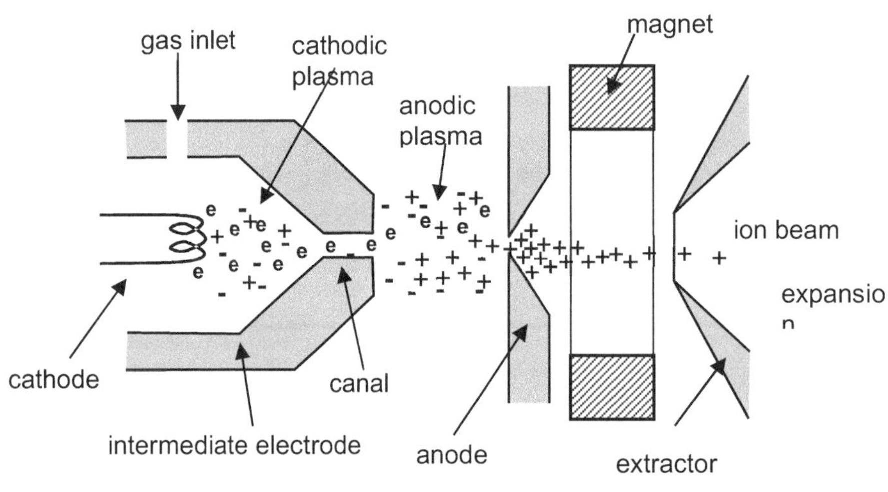

FIG. 3. Schematic of a Duoplasmatron ion source.

The physical processes in the duoplasmatron ion source are complex. When the cathode is heated, electrons are produced by thermionic emission and are accelerated by the electric field between the cathode and the intermediate electrode. Positive ions are produced by collisions between the electrons and gas molecules and by thermal dissociation. The positive ion yield is dependent on the gas pressure, which influences the collision frequency and electron energy. If the gas pressure is very low, the collision frequency is low, so the resulting positive ion yield is low. If the gas pressure is very high, the collision frequency is also high, but the electron energy is degraded and can be insufficient to induce significant molecular dissociation and ionization, so the positive ion yield is low.

The magnetic field in the second chamber focuses the anodic plasma allowing it to pass through the small anode aperture. The location of the magnet varies in different source designs, but effectively it is used to assist with beam constriction. The focused plasma also increases the thermal dissociation process.

Plasma that passes through the anode aperture into the expansion region forms a plasma bulb. The voltage present on the extractor electrode influences the surface of the bulb and positive ions which are present on its surface, will be extracted with the electrons being repelled back into the plasma by the negative potential. The form of the plasma surface is dependent on the arc condition, the magnetic field, and the extraction potential. Normally the extraction potential is constant, and it is necessary to change arc condition, magnetic field, or gas pressure to obtain maximum output.

The duoplasmatron ion source can be operated either in positive or negative extraction mode. When used in positive extraction mode on the tandem accelerator system, it is coupled to a charge exchange system to convert the positive ions into negative ions. The charge exchange system is mounted after the extraction potential and expansion chamber and is typically positioned between two Einzel lenses located before the low energy injector magnet.

4.4.2 Negative extraction with the duoplasmatron

By slight modification of a few parts constituting the ion source, it is possible to convert the duoplasmatron for negative extraction operation. Increased concentration of negative ions can be found on the edges of the plasma field in the anodic plasma region. By shifting the anode aperture offset from the centre axis it is possible to extract these negative ions from the plasma.

Usually, the manufacturer provides an additional intermediate electrode to be used for negative extraction operation that already has its aperture offset from the centre axis. This is necessary from a design point of view, since if the anode aperture were be shifted off axis, the entire beam optics would collapse afterwards. It is thus more appropriate to shift the entire source head offset by a small amount in contrast to anode aperture to achieve the desired result. The offset of the source head is typically similar in dimensions as is the extraction aperture installed in the anode. Additionally, to increase the negative ion extraction efficiency, the separately provided intermediate electrode has a special aperture shape that increases ion species discrimination before the anode aperture. It is typically constructed of a crescent shape design that covers the anode aperture in the centre region, viewed along the longitudinal axis, as shown in Fig. 4. To further improve the negative ion output current, the extraction aperture insert within the anode can be replaced with a larger diameter one. Manufacturers usually provide extraction apertures with various diameters for different operation modes with different gases used, with a design that ensures the correct pressure gradients within the ion source and beyond the extractor.

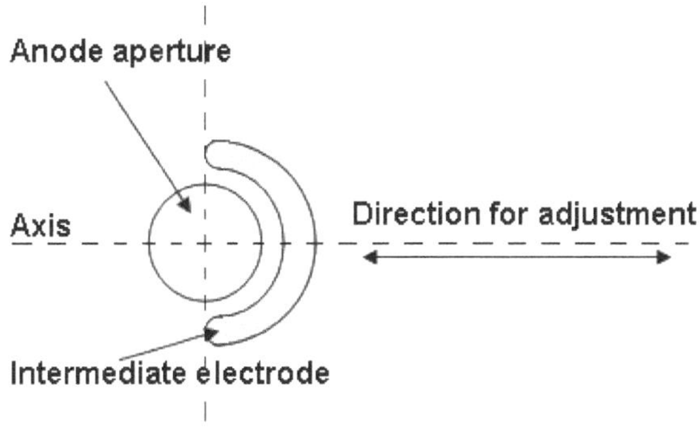

FIG. 4. Adjusting the intermediate electrode's radial distance from the anode aperture.

To optimize the off-axis displacement of the ion source head, it is necessary to measure the ion source current parameters. One needs to observe the actual ion current in the Faraday cup and the extraction current of the ion source. This is necessary because the ion source is operating in the negative extraction mode and a significant number of electrons are extracted along with the negative ions. The typical procedure is to obtain the maximum current of the negative ions in the Faraday cup while minimizing the extraction current. The extraction current is typically between one to two orders of magnitude larger and is therefore a good measure for extracted electrons. When adjusting the off-axis displacement for maximum negative ion output, there is a distinct turning point where only the extraction current starts to increase, and the extracted ion current remains approximately the same. At this point, only the electron extraction increases further, and this point marks the optimum displacement of the ion source head. Further increasing the displacement does not improve the extracted ion beam current. The extracted electron current needs to be kept as low as possible to minimize the electron bombardment of the exit cross field analysing magnet leading to possible damage of the latter. Excessive electron current also causes low energy radiation in the proximity of the ion source.

4.4.3 Faults and possible causes

Table 5 lists potential faults, the possible causes and rectification of a duoplasmatron ion source. The information is a guide only and can be combined with manufacturer's recommendations to rectify faults.

TABLE 5. LIKELY FAULTS, POSSIBLE CAUSES, AND SUGGESTED REMEDIES FOR A DUOPLASMATRON ION SOURCE

Fault	Possible Cause And Rectification
No arc current	Cathode 'cut-off'. Remove cathode and check quantity.
	No cathode current. Verify the power supply is working correctly.
	No arc potential. Verify the power supply is working correctly.
	No gas pressure or gas pressure too high. Verify gas circuit is operating correctly. Depleted supply or faulty regulator.
	Filament burnt out. Replace filament.
	Cathode shorting to plasma cup. Clean plasma cup.
No beam output	Anodic aperture is closed. Open aperture by scraping with a needle.
	No extraction potential. Verify the power supply is working correctly.
	Loss of arc. Gas pressure incorrect. Increase gas until arc strikes.
Low beam output	Cathode is not heated sufficiently. Increase cathode current.
	Gas pressure is not correct. Readjust and optimize gas pressure. Check gas supply.
	The anodic plasma is not focused on the anode aperture. Regulate the magnetic focus while observing the output.
	The anodic temperature is increasing. Verify the operation of the cooling system.
	Charge exchange material is low. Replenish charge exchange material.
	Charge exchange canal blocked. Clean canal and replace with new charge exchange material.
	Low current through filament. Replace filament.
	Poor vacuum. Check vacuum system

4.4.4 Servicing

The filaments used in duoplasmatron ion sources can have their life extended by coating them with a proprietary cathode coating media such as (Ba-Sr-Ca)CO_3. Manufacturers' manuals provide information on recommended methods for coating the filament. Typically, the coating procedure is as follows.

The filament is first prepared by washing in 20% nitric acid and distilled water and then rinsed with clean water. It is then blown dry with a hot air gun. The filament is then dipped into the coating solution up to about 10 mm from the filament supports. After removing from the coating solution, the filament is dried with a hot air gun. The coating and drying steps are repeated about 3 times to build up a thick layer. When completed, the coated filament needs to be mounted back inside of the ion source and the ion source evacuated to reduce the risk of contamination. The filament will require cautious outgassing and conditioning before use at normal operating conditions. The filament current is slowly increased over several hours, slowly increasing its temperature until ultimately the carbonated coating material is converted into an oxide. Duoplasmatron manufacturers provide recommended procedures in their operating manuals.

Lithium charge exchanges may need to be recharged every 4 to 5 weeks if they are in constant use. There is a tendency for lithium to migrate into the ion source internals, and as well towards the pre-accelerator tube, causing an apparent high consumption rate from the lithium reservoir. It will be necessary to clean the pre-accelerator at regular intervals, or at least annually. The pre-accelerator leakage current will increase as it becomes contaminated with lithium, so the operator can use this to indicate the condition of the pre-accelerator tube and set maintenance schedules accordingly, to mitigate this effect.

Refillable high pressure source gas cylinders can be adapted for use in any gas consuming ion source. This negates the need for commercially supplied pre-filled gas cylinders such as lecture bottles. To fill the gas cylinder, an adaptor can be made that allows the gas cylinder to be evacuated and backfilled with the desired gas. Care needs to be taken to select and use only high purity gases. When evacuating the ion source gas feed cylinders, it is very important that a cold trapped vacuum pumping system be used, with vacuum pumping maintained until a pressure of less than 0.1 Pa is achieved.

4.5 MULTICUSP ION SOURCE

Multicusp ion sources are high brightness ion sources, having a normalized proton beam brightness that can exceed 10, making them well suited for microprobe applications on tandem accelerators. It

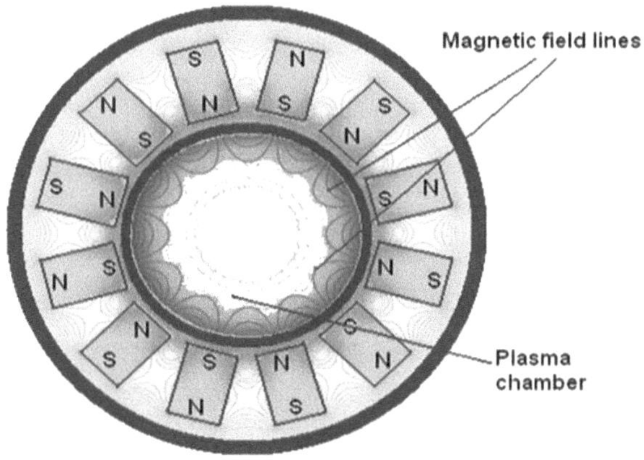

FIG. 5. Cross-section schematic of the plasma chamber with radially aligned multicusp magnets, with representative magnetic field lines inside the chamber.

utilizes a distinct shape of the magnetic field inside the plasma chamber, termed a multicusp field, as shown in Fig. 5. There are a variety of design types of this ion source that differ in the way the ions are produced. The most common type is a filament driven volume ion source, followed by an RF driven volume ion source and surface conversion type [16].

A multicusp ion source can produce both positive and negative ions depending on the specific design and can be used on all types of accelerator, from cyclotrons to low energy electrostatic accelerators. Since the low energy tandem accelerators require the injection of negative ions, the focus here will be on the negative ion (mainly H^- production) multicusp ion source.

The extracted proton beam current from a multicusp ion source can have a large dynamic range, ranging from nanoamperes up to hundreds of microamperes by changing the diameter of the extraction aperture, and even a few milliamperes with very large diameters. A multicusp ion source can also be utilized for the extraction of positive helium ions. On a tandem accelerator, where it is used to produce positive helium ions, the ion source is equipped with an alkali charge exchange canal to convert the positive helium ions to negative for injection into the tandem.

4.5.1 Principle of operation

The operating principle of a filament driven volume multicusp ion source is the production of vibrationally excited hydrogen molecules via various electron collision effects, which decay into negative hydrogen ions and neutral hydrogen atoms, as shown in Fig. 6.

When a hydrogen molecule travels through the plasma chamber it is vibrationally excited by a collision with a fast electron (hot electron) emitted from and near the hot filament. The vibrationally excited hydrogen molecule then drifts towards the colder region of the plasma chamber and interacts with one of the slower (cold) electrons which can attach to the hydrogen molecule. With another collision with the one of the slower electrons, the molecule then dissociates into one neutral hydrogen atom, and one negative hydrogen ion which is extracted from the plasma chamber. The extracted negative hydrogen ions constitute the negative ion beam.

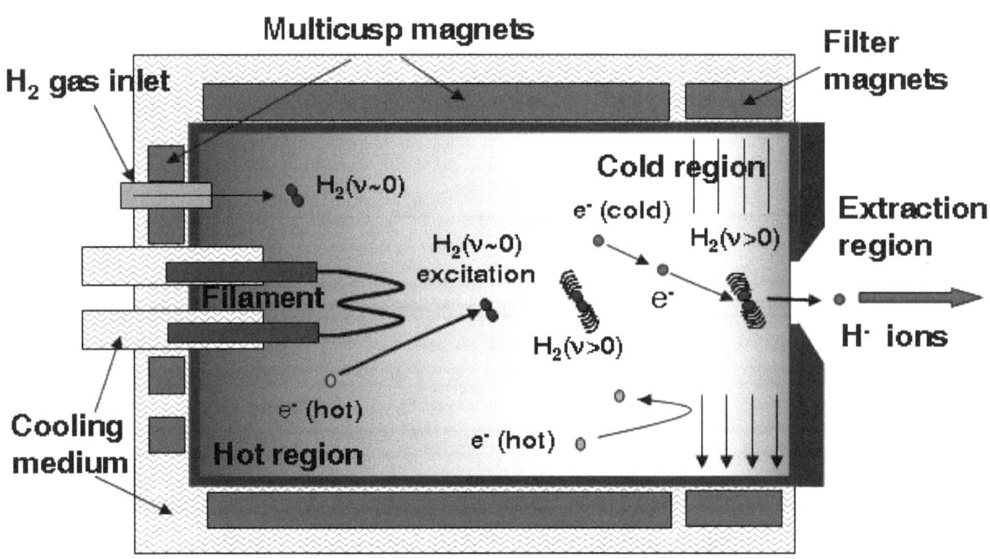

FIG. 6. Filament volume driven multicusp ion source schematics with major components with processes occurring marked.

Although it is possible to adjust the output current of the multicusp ion source over a wide range, the highest brightness can only be obtained in a specific output range. This is governed by the actual design of the ion source itself and depends on the shape and size of the plasma chamber, filament thermal output and electron emission, cooling temperature, gas pressure, electrode voltages, etc. To operate the multicusp ion source in an optimal way in a required output range, it needs to be designed and engineered with purpose selected characteristics. Operating the ion source for use outside of its optimal operating parameters shortens its service life and puts additional strain on supporting systems and components.

To better match the plasma characteristics inside the ion source with the object slit and accelerator, it is suggested to downsize the extraction aperture. Downsizing the aperture size helps reduce undesired and effects at the high energy side of the accelerator caused by protons near the edge of the extracted beam. Most of the beam's brightness is located at the core of the extracted beam and is not significantly affected by aperture downsizing. A smaller extraction aperture implies selecting a different setting on the plasma gas pressure, and it is necessary to check whether the gas pressure control valve can operate in the new gas flow range.

Using very high power beams can quickly erode beam defining object slits and apertures, produce high radiation fields, and locally melt beamline components made of stainless steel potentially causing significant damage to equipment. For example, protons with a beam current of 200 μA at 3 MeV (600 W power) can be deposited on an area of approximately 1 mm^2, when a good beam spot focus is achieved. A beam with such a high power density can quickly melt through any material. Therefore, it is essential to avoid manipulating beam diagnostic components and optimizing beam transport components using high ion source currents. Furthermore, when operating at high currents, the failure of a beamline steering component, such as switching magnet failure, the high power ion beam can penetrate its vacuum box creating a system vacuum failure. To protect the accelerator system, vacuum pressure gauges and interlocks need to be set and installed at critical locations.

4.5.2 Multicusp ion source preventative maintenance

Preventative maintenance for multicusp ion sources includes the following:

- Keep all surfaces and components clean and dust free. This helps to prevent electrical discharges over insulating surfaces and insulator breakdown.
- Always have a sufficient supply of ion source coolant at hand and regularly check the coolant levels. In the case of using deionized water as a coolant, always check that its conductance does not exceed the recommended value.
- Periodically check the coolant temperature when operating the ion source to ensure it is within the recommended limits.
- Always have various filters and replacement parts for the cooling system in stock;
- Periodically check that the coolant flow rate to be within the recommended limits and check for flow cavitation effects by listening for the possible occurrence of flow noise, and adjust the flow rate if necessary. If readjusting the flow rate puts its flow rate outside the operating limits, this can indicate possible residue buildup in the coolant lines and a complete dismantling of the system and general overall cleaning is urgently needed.
- Ion source coolant flow rate has a direct impact on the plasma behaviour, as the ion source head temperature affects the cool region of the plasma for electron attachment and, it also cools the permanent magnets of the multicusp field. Too high a temperature can cause magnetic field degradation which will lower the output current of the ion source.
- Periodically check for voltage discharges (mainly extraction voltage) by listening for the distinct noise it makes when the ion source is in operation. It is possible for corona discharge

to occur on exposed screws on and within insulator flanges. This will lead to insulator replacement in case of insulator breakdown. Significant extraction voltage current increase over time is usually symptomatic of insulator breakdown or surface contamination.

— When the filament is near the end of its operation cycle it requires replacement. Always have enough replacements available in stock. When replacing the filament, the main plasma chamber requires a thorough cleaning with fine grit sandpaper, and additional parts inside need to be checked for any material degradation and cleaned/replaced as necessary.
— When completely dismantling the ion source head for a general overhaul or major maintenance or cleaning, it is essential to correctly install the electron suppressor back to its original location and with the correct field orientation. It is usually placed off-centre, and its north-south alignments need to be checked to ensure its correct orientation.
— Periodically check the steering magnet for correct function as it defines the beam direction for injection into the accelerator. Check for any electrical discharges or unstable current/voltage output and replace/repair if necessary.
— Periodically check the gas supply system for proper function and check the gas quantity for an adequate reserve capacity.

4.5.3 Faults, possible causes, and rectification

Table 6 lists commonly encountered faults and their possible causes of a multicusp ion source. The information is a guide only and can supplement the manufacturer's recommendations for diagnostic methods to rectify faults commonly found.

TABLE 6. LIKELY FAULTS, POSSIBLE CAUSES, AND SUGGESTED REMEDIES FOR A MULTICUSP ION SOURCE

Fault	Possible Cause And Rectification
Plasma is oscillating	Lower the gas pressure in the plasma chamber or lower the filament current, or both.
Plasma does not start	Increase the filament current. Check the power supply output. Incorrect gas pressure. Change gas pressure until arc strikes.
Output current too low	Increase the anode current or change the extraction region within the plasma chamber by adjusting the first electrode and secondary electrodes, filament current, and gas pressure. Gas pressure is incorrect. Adjust and optimize gas pressure. Check gas supply. Poor vacuum. Check vacuum system.
Beam brightness too low	Change the plasma meniscus by adjusting the first and secondary electrodes, gas pressure, and filament temperature.
Filament current too low	Change filament. Check filament power supply.
Filament voltage too low	Clean insulating surfaces on filament feedthrough. Check filament power supply.
Anode current too low	Clean insulating surfaces/gaps between filament and plasma chamber housing.
Anode voltage too high	Increase filament current. Increase gas pressure.

4.6 CHARGE EXCHANGE SYSTEMS

The process for producing negative ions by passing a positive ion beam through an alkali metal vapour has been known for more than 50 years [17]. Commercially available negative ion sources for tandem accelerators can now reliably generate many microamperes of negative ion beams. The working principle of producing negative helium ion beams on small electrostatic accelerator systems is described below. The process is the same for producing negative hydrogen ions and other negative ion species.

4.6.1 Principle of operation

The starting point is to generate a beam of positive helium ions. Positive ion sources have been described in previous sections. The generated positive helium ions are input to a chamber containing an alkali metal vapour. Collisions between a positive helium ion and alkali metal atoms produce negative helium ions in a two-step process:

Collision 1: $He^+ + M \rightarrow He(1s, 2s) + M^+$ or $He^+ + M \rightarrow He + M^+$

In this first collision, the alkali metal atom (M) can transfer an electron to the positive helium ion forming a neutral helium atom in a metastable state. The neutral metastable helium atom has one electron in the 1s state and the other in the 2s state. Alternately, this collision can transfer an electron to the positive helium ion neutralizing it from further reactions.

Collision 2: $He(1s, 2s) + M \rightarrow He^- + M^+$

In this second collision, another alkali metal atom transfers an electron to the metastable neutral helium atom creating a negative helium ion.

Thus, the key to this process is the formation of a metastable neutral atom. The efficiency of this charge exchange process, the ratio of negative ions produced to positive ions input is low, reaching only about 2% in the best cases. The ratio depends on the species of negative ion to be produced, its kinetic energy, and the alkali metal used. Nevertheless, sufficient currents of negative hydrogen, helium, and heavy ions can be produced for many analytical applications on tandem type accelerators.

4.6.2 Charge exchange chamber

A charge exchange ion source consists of a charge exchange chamber coupled to a positive extraction ion source. Figure 7 shows a typical system for a RF negative ion source. The positive ions from the RF ion source enter the charge exchange chamber in which a fraction is converted to negative ions as described above.

An extraction lens at the exit of the charge exchange chamber collects and accelerates the negative ions away from the chamber and inputs to an Einzel lens to provide a focusing action for further beam transport.

Typical charge exchange systems use either rubidium, sodium, or lithium as the alkali metal. The melting points of these three alkali metals are approximately 39°C, 98°C and 180°C respectively. The design of the charge exchange system is mainly governed by the chosen alkali metal, its charge exchange efficiency, and the complexity of the maintenance procedures. During the development of charge exchange technology, the alkali metals lithium, potassium, sodium, rubidium, and caesium were investigated for their charge exchange effectiveness. The conversion yields for He^- as a function of the He^+ ion beam energy is presented in Ref. [18] and shows that rubidium produces the highest He^- yield.

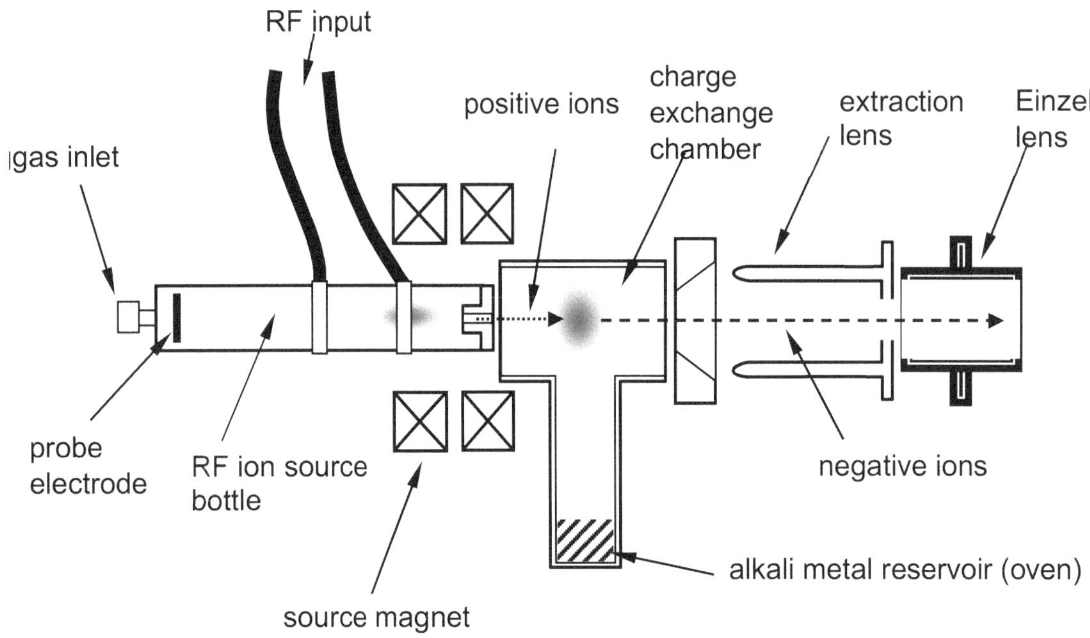

FIG. 7. Charge exchange ion source with an RF ion source as the source of ions.

Commercial negative ion sources using lithium or sodium as the alkali metal are in many cases, now being used instead of rubidium. Lithium and sodium, being the least reactive of all the alkali metals, are considered less hazardous to handle than the more highly reactive rubidium. For accelerator operators, an additional benefit is that the price of lithium or sodium metal is significantly less than rubidium.

The alkali metal vapour is produced in a heated reservoir mounted on the charge exchange vacuum chamber. The temperature maintained in the oven is critical, for rubidium it is typically 220–260ºC and for lithium it is usually 480–580ºC. The amount of vapour produced for charge exchange in the chamber is partially regulated by condensing some of the vapour onto the cooled wall surfaces where it eventually flows back down into the oven and the cycle begins again. The condensing rate of the vapour is crucial, and directly influences the effectiveness of charge exchange. The temperature of the charge exchange chamber needs to be monitored and preferably regulated. For rubidium it is usually maintained between 50ºC and 60ºC, although this may vary in some sources, and for lithium it is near or above 200ºC. It is also possible that the charge exchange chamber is designed in such a way that it maintains optimal temperature gradients for the alkali metal vapour condensation on the walls of the chamber by utilizing built-in barriers, and compartments that enable an optimal thermal operating regime.

A turbomolecular vacuum pumping system is usually the most efficient type for these ion sources, providing a base pressure typically less than 0.1 mPa. If helium is used as the feed gas in the positive ion source, then cryogenic pumps are not appropriate, as helium cannot be effectively pumped by cryogenic vacuum pumps, as it saturates the condensers and causes the vacuum pressure to rise to unacceptable levels.

4.6.3 Faults and possible causes

Table 7 lists potential faults and their possible causes. Many of the faults that occur in this type of source are common with those of the RF positive ion source described previously. The information is a guide only and can be combined with manufacturer's recommendations to rectify faults.

TABLE 7. FAULTS RELATED TO A TYPICAL CHARGE EXCHANGE ION SOURCE AND POSSIBLE CAUSES

Fault	Possible Causes
No beam output	No beam measurement due to faulty Faraday cup.
	No beam measurement due to faulty microammeter or input cable.
	Loss of gas feed due to fault in metering valve or depleted gas supply.
	Loss of probe, extraction or focusing high voltage.
	Loss of the RF power due to oscillator failure or output triode failure.
	Loss of charge exchange process due to low alkali metal vapour flow.
	Low vapour flow due to insufficient alkali metal in oven, faulty heater, or incorrect oven temperature.
	Chamber temperature too high.
Low beam output	Inaccurate beam measurement due to faulty Faraday cup.
	Inaccurate beam measurement due to faulty microammeter.
	Dirty or contaminated Pyrex glass bottle.
	Extraction canal eroded or partly blocked.
	Low output from RF triodes.
	Insufficient gas feed.
	Loss of focus of the ion beam due to breakdown of extraction voltages breakdown of Einzel lens voltages, or breakdown of pre-acceleration voltages.
	Low charge exchange efficiency due to low alkali metal vapour flow.
	Low vapour flow due to insufficient alkali metal in oven, faulty heater, or incorrect oven temperature.
	Poor beam transmission due to alkali metal vapour condensed near extraction aperture.
	Alignment incorrect.
Unstable beam output	As for low beam output.

4.6.4 Charge exchange efficiency

To check the effectiveness of the charge exchange system, there are a few possibilities to do so, and they depend on the specific design of the ion source/charge exchange system setup. The procedure differs depending on whether there is an RF ion source coupled to a rubidium oven, duoplasmatron ion source to a lithium oven, or multicusp ion source to a sodium oven. But in all cases the procedure involves measuring and comparing the positive ion extraction current and negative extraction current. The positive ion current is measured on the Faraday cup after the charge exchange system and before the accelerator without the charge exchange system in operation. This is measured either on the Faraday cup on the pre-accelerator, after the injection magnet, etc. depending on the specific design of the low energy injection system. Then, the charge exchange system is put into operation and the output current is again measured on the same Faraday cup. By varying and optimizing the ion source parameters, beam optics and charge exchange oven temperature, the maximum output current is obtained. The comparison between the two currents, positive and negative, then yields the charge exchange effectiveness. Logbook records of such measurements over an extended history can provide a valuable aid to optimizing the performance of the ion source, and in fault finding.

4.6.5 Cleaning of alkali metal from fixed ion source parts

Caution is needed to remove any alkali metal that remains on ion source components after the oven has been removed. The following information is provided as a guide for cleaning the ion source without the need to dismantle it, prior to recharging the reservoir. It requires the person undertaking the operation to wear appropriate protective clothing, including thick rubber gloves and eye protection.

To clean a rubidium charge exchange system, the following procedure can be used:

— First, warm the charge exchange system to liquefy any rubidium that has solidified inside. A large, clean, and empty container such as a 60-litre garbage bin is filled with water to a depth of about 50 mm and placed directly underneath the reservoir opening of the ion source assembly to catch any droplets of rubidium as the reservoir is removed. Before removing the oven/reservoir, argon needs to be flowing through the ion source to maintain an inert atmosphere, preventing residual rubidium in the chamber from oxidising.
— As the source is disassembled, slightly water damped cotton buds are used to dab away rubidium from the source parts. It is normal to hear a cracking sound when the rubidium reacts with the water. The cotton buds can be quickly dropped into the container of water if they ignite. Any rubidium that falls from the ion source during the disassembly process will be captured in the container and neutralized by spontaneous combustion.
— The disassembled source parts need to be immediately, and cautiously immersed in water to neutralize any remaining rubidium before cleaning.

For lithium and sodium charge exchange systems, the much higher melting temperatures of these alkali metals makes it prohibitive to heat the system to liquefy any solid remnants inside. The disassembled source parts need to be cautiously immersed in water to remove any solid remnant alkali metal before cleaning.

4.7 CLEANING OF ION SOURCE COMPONENTS

Ion source components need cleaning after an extended period of operation. The cleaning process is necessary to remove any deposits that may cause electrical breakdown of insulators, contamination of surfaces leading to high recombination rates or contaminated plasma, high stress points encouraging electrical discharge, and metal deposits that modify the electric fields or block apertures. The need for cleaning will be typically indicated by a lower yield from the ion source, an unstable output, or a contaminated beam. The operators of the ion source need to develop an understanding of the frequency of cleaning that is necessary to maintain an effective output. This will be gained through operational experience and needs to be incorporated into the routine scheduled maintenance.

4.7.1 Insulators

Several types of insulator material are used in ion source constructions. Hard materials such as ceramics and glass, and softer materials such as plastics and boron nitride are all used in various situations. A cleaning method needs to be selected based on the component structure and material, and the risk of damage by the cleaning agents and actions. Irreparable damage to insulators may result from selecting incompatible cleaning agents and actions, so it is important to always use the least physically and chemically aggressive techniques. The highly polished surface on some insulators need to be preserved; therefore, whatever cleaning method is selected, it has to not scratch or scuff the surfaces. Penetrating cleaning chemicals need to be avoided, as they might not be fully removed from the surface, leaving a potentially conductive path for electrical discharges.

Warm soapy water used with a soft scour pad is a commonly used cleaning method. This is particularly effective for any parts that may have traces of caesium or rubidium residues on the surface. Rinsing in ethanol assists in removing water and when followed by acetone, is an effective cleaning method, but it may soften many plastic type insulators. After cleaning, insulators need to be thoroughly dried before reassembly.

Where manual washing procedures are not possible, for example, deep inside the body of an electrical feedthrough, cleaning with a fine jet abrasive compound is effective. However, all traces of abrasive need to be removed from the component using a high pressure blast of compressed air or dry nitrogen.

4.7.2 Metal components

Before cleaning metal parts, it is important that they first be assessed whether they may sustain damage from being immersed in a cleaning agent, including water. Some detergents are acidic and may oxidize the surface or leave a residual surface film. If uncertain, the cleaning agent needs to be first tested on a small area of the metal component.

An efficient method of cleaning most metal parts is in warm soapy water using a scouring pad, finishing with a thorough rinse in warm water. The components need to be dried with compressed air or dry nitrogen. Finally, they need to be wiped with acetone and lint free cloth. Parts that will contact alkali metals in their final assembly can be vacuum or oven dried, if available.

If the deposits to be removed are persistent and not easily removed by the above method, an abrasive 'wet and dry' paper, grade P1200, can be used either by hand to polish the component, or mounting the component in the chuck of a lathe and polishing the piece as it slowly rotates. In either method, the polishing is best achieved with water. For very persistent deposits, localized rubbing with a coarser grade abrasive paper, such as grade P800, can be carried out followed by polishing with the finer P1200 grade.

4.7.3 Ionizers

Helical and spherical ionizers can be cleaned using a fine jet abrasive. After cleaning, any abrasive compound or dust remaining on the ionizer needs to be removed with a high pressure jet of compressed air or dry nitrogen. It is important to avoid blasting the electrical and thermal insulating material from the ionizer shell around the leads. The ionizer surface has not to be touched with a bare hand after cleaning and is best stored in a plastic bag until ready for use.

4.8 ION SOURCE GAS PRESSURE REGULATION

The output of an ion source depends on the pressure of the discharge chamber. The regulation of this pressure is achieved by controlling the flow of the incoming feed gas. The commonly used regulation systems are:

— Palladium leak: it is used exclusively for hydrogen, using the property of molecular hydrogen to dissociate on one side of a Pd foil and then to migrate to the other side. By controlling the temperature of the foil, the flow can be controlled. This technique has the advantage of purifying the hydrogen gas that is reaching the discharge chamber.
— Thermo-mechanical leak: it consists of an invar (Ni-Fe alloy) rod inside a stainless steel cylinder that is connected to a ball valve. Invar has a different thermal coefficient of expansion than stainless steel, and by changing the system temperature, the gap of the valve is changed

and so the flow rate changes. This system requires long stabilization times (hours) and small changes of the temperature will affect the flow rate.

- Precision needle valve: A soft metal pad is compressed against a harder metal seat, creating the necessary seal for precise gas flow control. These valves are typically manually controlled. They can have undesirable backlash issues.
- Mass flow controllers: They consist of a sensor, a bypass, a flow rate control valve, and a special sensing circuitry. The sensor measures the mass flow rate of the gas, and the flow rate control valve modifies the flow rate so that the difference between the measured flow rate and a manually set flow rate is zero. These devices have the advantage of a fast and stable flow gas regulation, resulting in an excellent operation for use in ion sources.

5. ACCELERATOR CHARGING AND CONTROL SYSTEMS

There are three main types of charging system used in low energy electrostatic accelerators:

- Belt charging Van de Graaff;
- Chain charging Van de Graaff;
- Cascade solid state multiplier.

Belt charging accelerators were used in electrostatic accelerators up to about 1990. This type of charging system is no longer in production and most accelerators worldwide that used this charging system are no longer operational. They have either been decommissioned or converted to chain charging. A description of belt charging is included in this publication to preserve knowledge about belt charging machines, and to acquaint the reader with their technology and principles, which have evolved to drive modern machines and control systems.

Robert Van de Graaff developed the belt charging technology, which was adapted to produce high voltages in the earliest accelerators. These accelerators became known as Van de Graaff type accelerators. The belts looked like a rubber conveyor belt and had sufficient electrical properties to carry ~1 mA of charging current to the high voltage terminal.

A mechanical charging system was developed by Ray Herb at the University of Wisconsin in the 1960s that consisted of a chain made up of stainless steel cylinders, coupled together with insulating nylon links. These chains could carry only up to 150 μA but with greatly improved charging stability. Multiple chains can be fitted to accelerators increasing their charging capacity.

In 1930, Cockroft and Walton described how to use a voltage multiplying circuit to drive a potential drop accelerator. This technique was improved on, modified, and developed into other similar technologies which in general are called cascade generators, due to the multiple series voltage generation sections.

5.1 BELT CHARGING SYSTEMS

Original charging belts consist of a cotton multi-layer carcass, which has been coated with a vulcanized rubber material on both sides. For accelerators that operated up to 1 MV, the width of the charging belt was 15 cm and for higher voltage machines, the belt width was 52 cm. Charging belts can cause variations in the up-charging current to the terminal due to surface inhomogeneities and non-uniformities. Vulcanizing of the belt was done in sections causing a thicker layer of rubber where the sections overlapped. Charge is transferred to the surface of the belt from a high voltage biased screen which scrapes along the belt's surface as it rotates. A similar screen in the terminal contacts the belt to remove the charge. Figure 8 illustrates a simple belt charging system, which is similar for both single and double-ended accelerators.

Modern belt technology using new materials has not been able to reproduce the same specifications as the original belts. Materials such as polyurethane on composite substrates have been used with some success, but problems with self-charging and abnormal heating have limited their use. Belt charged accelerators suffer a problem with belt dust generated by the screens scraping the surface fast moving belt. Over time, the accumulation of belt dust over the accelerator's internal components leads to terminal voltage instabilities. The belt dust needs to be removed during every tank opening.

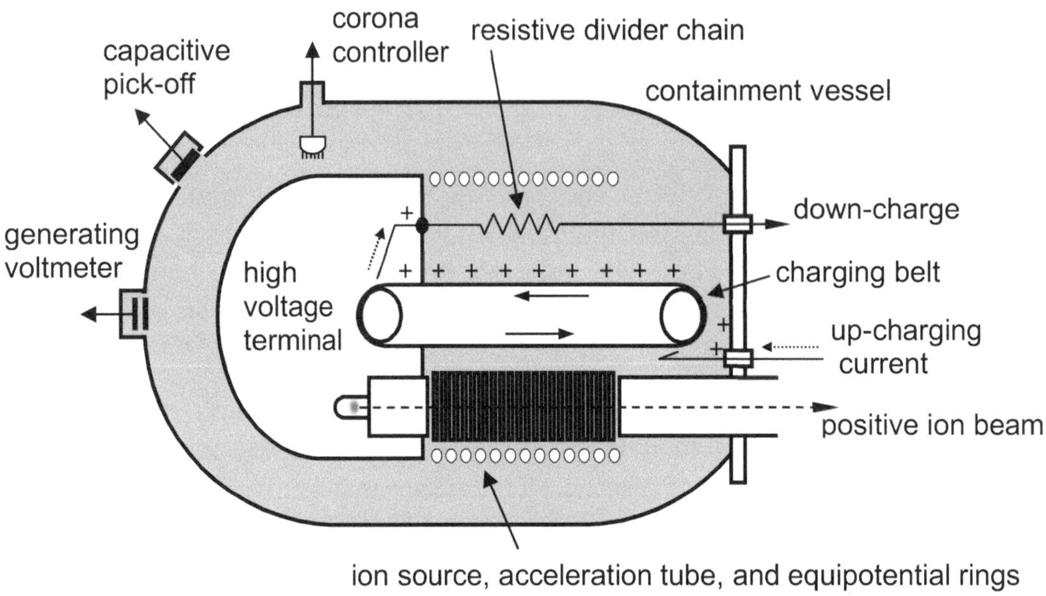

FIG. 8. A belt charging and control feedback system in a single ended high voltage generator.

The voltage that is needed to induce sufficient charging current on the surface of the belt is typically 5–10 kV DC. The charging and collector screens cannot always maintain good contact with the belts' surface as it moves past due to variations in the belt's thickness. This results in variations in the amount of charge applied to and removed from the belt. The voltage on the terminal can thus show a voltage ripple of several kV peak-to-peak, with a repetitive pattern corresponding to a complete revolution of the belt. This can be seen when observing the output from a capacitive pick-off (CPO) located on the inner wall of the accelerator (see Section 5.5.4).

5.2 CHAIN CHARGING SYSTEMS

Pelletron chain charging systems produce minimal fluctuations in terminal potential because the charge is deposited on the chain's pellets by induction, a process insensitive to surface irregularities which is a common problem with belts. Pelletrons run dust free and need to be replaced only once every 5 to 7 years. The principle of the induction type chain charging system is illustrated in Fig. 9.

As the grounded drive pulley rotates, the pellets in contact with the pulley pass a negatively charged inductor electrode that pushes electrons off the pellet. As the pellets move through the inductor field away from the pulley they become positively charged. The charge remains on the chain pellets as they travel towards the terminal. When the charged pellets reach the terminal, they pass through a negatively biased suppressor electrode, which prevents arcing as the pellets contact the terminal pulley. The pellets that have passed the suppressor electrode field lose their positive charge to the terminal pulley, which then accumulate on the terminal. Chain charging systems can have both down-charging and up-charging. Down-charging works identically to up-charging, except the inductor/suppressor polarities are reversed, and it effectively doubles the charging capacity of the chain. Some systems are configured with two small, slightly conductive 'pick-off' pulleys; one that contacts the chain entering, and the other leaving the terminal pulley. The pick-off pulleys provide the voltages for the terminal suppressor and inductor electrodes by drawing a tiny amount of charge

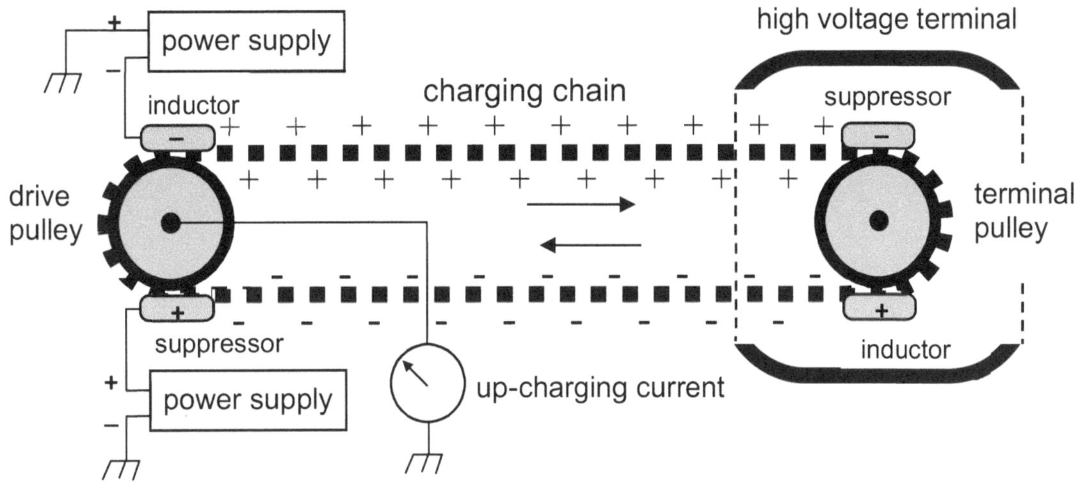

FIG. 9. The principle of the chain charging system used in both double and single-ended accelerators.

from the chain. Each pulley therefore biases the opposing electrode eliminating the need for a high voltage power supply in the terminal to produce a negative charge for down-charging.

The chain system can carry charging currents of 100–200 μA or more per chain to the high voltage terminal. The drive pulleys are typically 30–60 cm in diameter and the motors are supported on movable platforms which are counterweighted, automatically providing proper chain tension. The charging current in chain charging systems is inherently more stable than in belt charging, as all the charging pellets are essentially identical in shape, size and current carrying capacity.

It is prudent to check the charging system during routine maintenance. A procedure is to first remove any grounding safety grounding straps from the terminal then connect a ~1 MΩ resistor between the terminal and ground and connect an oscilloscope across the resistor to show voltage. Then the chain drive motor is started and a low charging current, typically not more than 100 μA, is applied. The oscilloscope display will show the terminal voltage and voltage ripples. The voltage ripples can have various frequencies: the transfer of charge due to the many individual pellets produces a high frequency, while slower moving components such as the pulleys produce lower frequencies. Typical voltage fluctuations are < 0.1%. Any anomalies observed can help identify potential problems before tank closure. The information on the oscilloscope display needs to be recorded and if possible, captured for future reference.

5.3 CASCADE CHARGING SYSTEMS

In 1932, John Cockroft and Ernest Walton developed a voltage multiplier generator (Fig. 10) based on the Greinacher voltage doubling circuit (outlined by dashed lines in Fig. 10). The input is an AC voltage of amplitude U_0. The charge is transferred stepwise from the AC source to the high voltage terminal in n identical stages, called cascades, producing a no-load DC terminal voltage of $2nU_0$. This circuit can produce voltages in the megavolt range. Under load, there is a voltage drop and a ripple voltage on the output. To keep these quantities small, the input frequency and capacitances need to be large, and the number of cascade stages small. Typical values are f = 0.5 kHz to 10 kHz, C = 1 nF to 10 nF, and n = 3–5 stages.

Circuits are now based on parallel-driven circuits, as shown in Fig. 11.

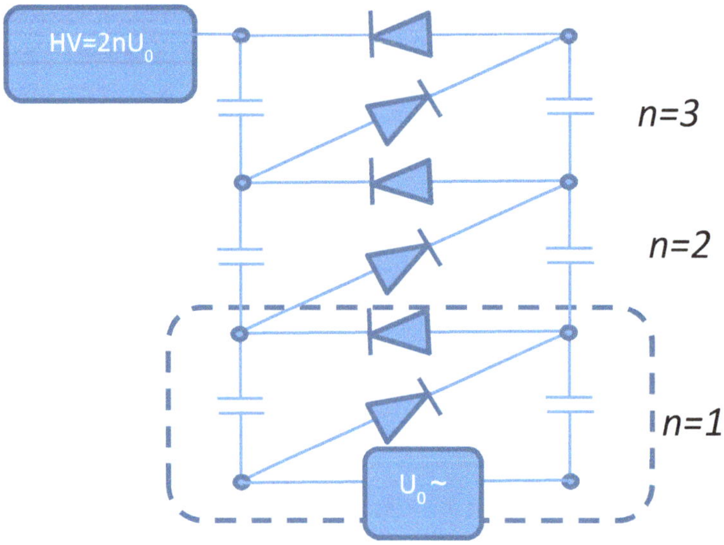

FIG. 10. Electrical circuit of a Cockcroft-Walton voltage multiplier. Dashes outline the Greinacher doubling voltage circuit.

The input is an AC voltage of amplitude U_0. The charge is transferred stepwise from the AC source to the high voltage terminal in *n* identical stages, called cascades, producing a no-load DC terminal voltage of $2nU_0$. This circuit can produce voltages in the megavolt range.

Two large semi-cylindrical RF electrodes (dynodes) surround the accelerator column. The dynodes are fed with the output of an LC resonating circuit, at frequencies of several tens of kHz inducing a voltage in the corona rings surrounding the accelerator column. Voltage multiplication is produced by coupling the corona segments to rectifiers, with each segment achieving DC voltages ~50 kV.

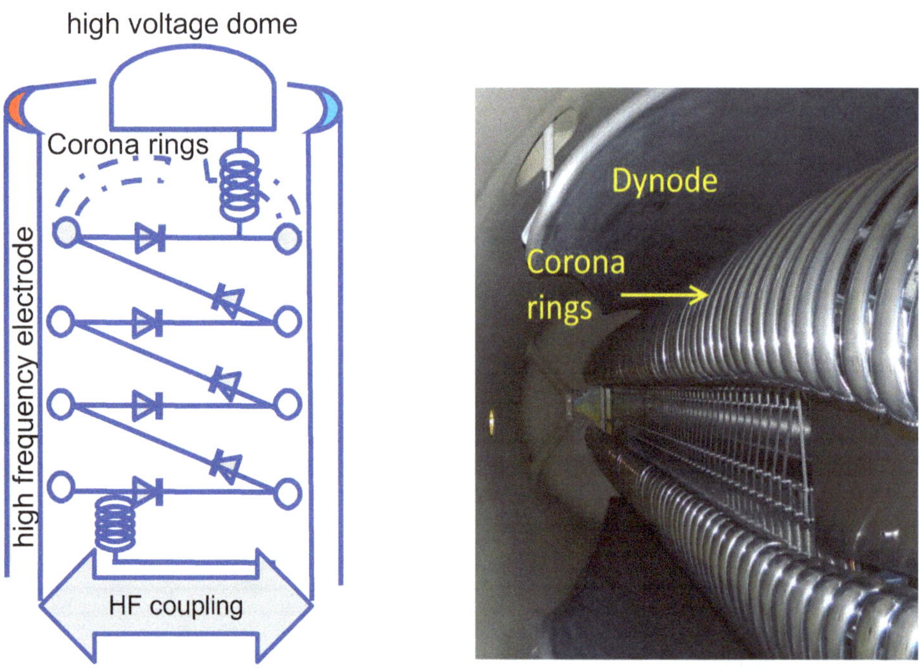

a) Schematic of a parallel-driven cascade generator.

b) actual construction

FIG. 11. Cascade charging system.

Terminal voltages of more than 6 MV can be attained. The terminal voltage stability is typically 0.1% or better and can provide up to 500 μA of beam current in cascade driven Tandetrons.

5.4 CHECKING THE PERFORMANCE OF CHARGING SYSTEMS

The charging system can be checked for performance in various ways. The primary way is to regularly monitor and/or log the baseline ripple of the terminal voltage during normal operation. Increased ripple can give an early warning of potential problems. Further investigation may be undertaken during tank-off maintenance when the internal components of the accelerator can be visually checked.

5.4.1 Belt charging

A belt charging system can be comprised of three components, each of which needs to be checked. They are:

— The charging power supplies;
— The charge transfer screens;
— The belt and belt guides.

5.4.1.1 Power supplies

The power supply is tested off-line, as follows:

— Disconnect the power supply from the accelerator's charging circuit and connect a suitable resistor from the power supply output to ground.
— Activate the power supply and increase the output current to the nominal operating value. If the current is stable to within specifications, then the power supply is operating correctly. A high-quality power supply has to regulate output current to at least 0.01%. Poor regulation could mean that the power supply itself is faulty, or the controls that drive it are faulty.

5.4.1.2 Charge transfer screens

Checks need to be made on the high voltage feedthrough on the tank, and the charging and collector screens. The resistance of the feedthrough needs to be greater than 2000 MΩ. This resistance can be checked by isolating the feedthrough, applying a high voltage (~5 kV DC) and measuring the current flowing from feedthrough's conductor to the accelerator tank.

The resistance between the charging screen and its feedthrough needs to be less than 10 Ω. The charging and terminal screens need to only contact the surface of the belt lightly; too lightly and there will be insufficient charge deposited on and removed from the belt, too heavily and they will scrape the surface excessively reducing the charge carrying efficiency and lifetime of the belt. The width of the charging screen needs to be slightly less than the belt width to prevent any accidental contact with or corona discharge to the drive motor pully which is at ground potential. The collector screen needs to be wider than the belt to ensure effective collection of all the charge on the belt.

5.4.1.3 Charging belt and belt guides

If an abnormally high charging current is required to reach and sustain the nominal terminal voltage, then charge loss is occurring. The belt guides need to be inspected for correct alignment or surface contamination that can cause a path for charge leakage, and then cleaned, replaced or adjusted, as necessary. To help locate the problem in hard-to-find situations, the terminal is connected to ground;

then, by connecting a microammeter between ground and various points on the column, any detected microammeter current will help narrow the search to a region where charge loss is occurring.

The worldwide inventory of original manufactured belts has been exhausted and they are no longer available. New types of replacement belt have been used that are relatively successful. A basic type that has had some success is a polyester reinforced polyamide belt with a polyurethane coating to give good wear properties. It is constructed as a standard conveyor type belt with a seamless seam, spliced in a 'Z' pattern. As polyamide is a softer material than the rubber used in original belts, some belt guides may have to be removed to accommodate the new belt to reduce the possibility of the belt scrapping against the guides and tearing.

5.4.2 Chain charging

5.4.2.1 Power supplies

If the terminal voltage is not stable, then either the external power supply is faulty or there is a fault inside the tank. The high voltage feedthroughs usually have resistors attached that serve to act as current limiter to protect the external power supply if spark occurs inside the tank. The resistors need to be checked to see that their resistance is within specifications. The resistance between the metal pellet contacting the drive pulley and the drive pulley needs to be very low, typically less than 20 Ω. A high resistance is indicative of a misalignment or excessive wear of the contact bands on the chain.

5.4.2.2 Charging chain

Chains need to be periodically examined for wear and tear. Each metal pellet is connected to its neighbour by a nylon link. The nylon link is attached to the metal pellet by rivet that, if fails or works loose, will likely result in the nylon link detaching causing the chain to break. All pellets and their rivets need to be thoroughly checked for tightness and corrosion and defective rivets replaced. Corrosion is typically indicated by a rust-coloured ring surrounding the rivet head. In the terminal, the adjustment of the gaps between the chain and the electrodes can be checked with a standard jig.

5.4.2.3 Charge pick-off and idler wheels

If idler wheels are installed, they need to be kept clean and their bearings running noise free. Noisy bearings are indicative of a worn or damaged bearing, as is also signs of lubricant leaking from the bearing. Lubricant leaking can create a pathway for some current to be lost from the charging chain and flow to the column. The resistance from the charge pick-off wheel to the terminal is normally about 1 MΩ and needs to be regularly checked.

5.4.3 Cascade charging

The failure of a cascade charging system usually results from the failure of an electronic component such as a transformer, a rectifier, a capacitor, or a resistor. As the components in a cascade power supply operate at high voltages, special precautions need to be taken to ensure safe testing and special test instruments may be required to test the components at or near their normal operating voltage.

Before servicing the high voltage power supply, the output needs to be grounded, and all high voltage capacitors need to be discharged to ground. Capacitors can retain significant charge even after the power supply is switched off. Discharging needs to be done with a current limiting resistor to avoid damaging arcs in case of a charged capacitor. While working on these types of charging system, it is important that a grounded braided copper strap is connected to the points where high voltages are produced or conducted to.

Troubleshooting the power supply begins with a visual inspection of the components and wiring. Components that have failed catastrophically will usually be apparent. If there is no output voltage, broken wiring or loose connectors may be the cause. Continuity can be checked with a multimeter. If all connections are satisfactory, then the power supply fuses need to be checked. All other tests will involve component testing as described below.

5.4.3.1 Testing the high voltage transformer

Testing the high voltage power supply begins with testing the high voltage transformer, as follows:

— Isolate the transformer from its secondary side so there is no load and measure the primary current versus primary voltage. The secondary voltage can be simultaneously measured with an AC high voltage meter.
— The primary current I_m (the magnetizing current) and the ratio of the secondary to primary coil voltage needs to be almost constant. During the test, I_m needs to have a linear response relative to the applied voltage. If it is non-linear, then there may be corona breakdown on the high voltage terminals or there may be a problem internally. If there is a short circuit between the windings, there may be an abnormal temperature rise in the transformer. If the high voltage transformer is found to be operating normally, test the rectifiers and the filter capacitors that make up the power stack.

5.4.3.2 Testing the high voltage rectifiers

The specifications for the high voltage rectifiers can be found in the power supply manual or in the manufacturer's data sheets. The rectifiers normally used are silicon, series connected, controlled avalanche diodes. A standard single silicon p-n junction rectifier can withstand 1–1.5 kV of reverse bias while a controlled avalanche diode may withstand up to 100–150 kV. Multiple rectifiers joined end to end in a stack arrangement are best tested together to save time and effort. The rectifier stack needs to be removed from the power stack prior to testing. If the measured forward voltage across a rectifier stack is more than 5% out of tolerance, as compared with a known value, then it needs to be dismantled, and each rectifier tested individually.

5.4.3.3 Testing forward characteristics

Testing is best done with a power supply of about 200–500 V DC, adjusted to have a constant current of 1 mA. The voltage is applied in the forward direction and the voltage drop measured. The measurement of a set of rectifiers needs to start from one end of the power stack in the accelerator through to the other. It is useful to plot the values to determine the state of the whole power stack. Rectifier stacks that have had rectifiers replaced need to be tested again and the measurement recorded. All measurements need to be kept for future reference.

5.4.3.4 Testing filter capacitors

The capacitance in a filter capacitor needs to be between +30% and –20% of its nominal value. It is important to monitor the capacitance values regularly, as a low capacitance may cause increased ripple in the output voltage. In high voltage capacitors, the stored energy is generally high, so a breakdown inside the device may cause damage to the outer contact of the capacitor.

There are two ways to determine if a capacitor is in good condition. The primary way is to directly measure the capacitance across the isolated terminals with a suitable measurement instrument. The other way, which is only suitable for persons having experience with high voltages, is to observe the discharge spark when a suitable high voltage resistor, adequately supported by an insulating rod, is placed across its terminals. A 'healthy' spark will discharge from a good capacitor. To ascertain what is 'healthy', it is best if a set of capacitors is checked and compared with each other. It is unlikely that all would be damaged so the best spark can become the benchmark.

5.4.3.5 Testing resistors

Resistors used in cascade charging systems are typically higher than 10 MΩ, so standard multimeters are not suitable for measuring their resistance. Instead, a high voltage 'Megger' type device is needed. The insulating supports and connectors for the resistors need to also be checked for conductive tracks or damage and replaced, as necessary.

5.4.3.6 Quality factor measurements of parallel driven cascade accelerators

Parallel driven cascade accelerators use an LC circuit operating at a resonant frequency to feed power into the diode cascade capacitors. In this circuit, a high amount of reactive power is present, in the MW range. The quality factor (Q-value) of the LC circuit gives the ratio of reactive power to resistive (dissipated) power. Since the resistive loss is mainly found in the resistance of the resonant coil, a quality factor of 10^3 implies that the coil from the circuit will dissipate 1 kW of heat for every 1 MW of reactive power.

The Q-value may drop due to damage in the resonant coil or rectifiers from the cascade stack. If the drop in Q-value increases, the circuit driver will not operate properly.

The measurement of the Q-value is typically performed by exciting the tank LC circuit with an external signal generator at the resonant frequency, to measure the loss resistance of the circuit. The Q-value is $Q = Z_{coil}/R_{loss}$, where Z_{coil} is the coil impedance, and R_{loss} has to be determined.

At the resonant frequency, all the current is dissipated to the loss equivalent resistance, giving

$Q = \omega \times L_{coil} \times I/V$, where ω is the resonant frequency, L_{coil} is the coil inductance (typically tens of mH). I is the current in the circuit (can be measured as a voltage drop in a series added resistor) and V is the voltage at the entry of the circuit.

Another way to measure the Q-value is by using an AC current probe (supplied by the manufacturer). The current probe is connected in series between the signal generator and the tank LC circuit. The Q-factor can be defined by the ratio $Q = \omega/3$ dB bandwidth. The resonant frequency is defined when the driving sinusoidal wave from the signal generator and the current readout from the current probe are in phase. A 3 dB bandwidth is achieved when the phase shift between is 45°. Both signals can be observed with an oscilloscope.

In the first step, the resonant frequency needs to be precisely defined by changing the signal generator frequency until the two signals have the phase shift of zero degrees. The frequency is then increased slightly to reach 45° phase shift and the frequency is recorded. The same procedure is conducted to note the frequency at 45° phase shift below the resonant frequency. The 3 dB bandwidth is defined as the difference between the upper and lower frequencies.

5.5 CONTROL SYSTEMS

The electrostatic accelerator develops a high positive potential voltage by the transfer of positive charge to an electrically isolated high voltage terminal. The terminal is supported by means of either insulating plastic panels in the case of small accelerators or, in larger machines, by insulating columns made of glass and metal glued together. To stabilize the charge on the terminal, the electric field between the terminal and the grounded end of the insulating column (and acceleration tube) is distributed uniformly by a series of high voltage resistors in series across each charging plane or insulating gap. These resistors are chosen to drain only a fraction of the available charging current from the terminal to maintain charge stability. The terminal potential is determined by applying a charge I_{up} such that $V_t = I_{up} \times R_{col}$.

For belt and chain type mechanical charging systems, there are always variations in the charging currents (I_{up}) delivered to the high voltage terminal (i.e. I_{up} is not constant). The charging current is time-dependent, having both short term fluctuations and long term drifts. The short term fluctuations are produced by the characteristics of the two mechanical charge transport systems, as previously described. The long term drifts arise from AC line changes during the day, changes in temperature of components, and general physical changes in components due to wear. Cascade type charging systems experience fewer short term fluctuations but are still prone to some long term drift. All these variations need to be controlled and regulated to maintain a high stability of the high voltage on the terminal.

Figure 12 shows a general layout of a tandem accelerator with a single charging chain, the components used to charge the terminal, and the control and feedback devices, which are discussed in the following subsections.

5.5.1 Corona stabilization

Corona stabilization provides some control over the amount of charge that is held by the terminal. It gives the operator the ability to maintain short term control over the terminal voltage and thus ion beam energy. Typically, it is desirable to regulate the terminal potential to maintain the stability of

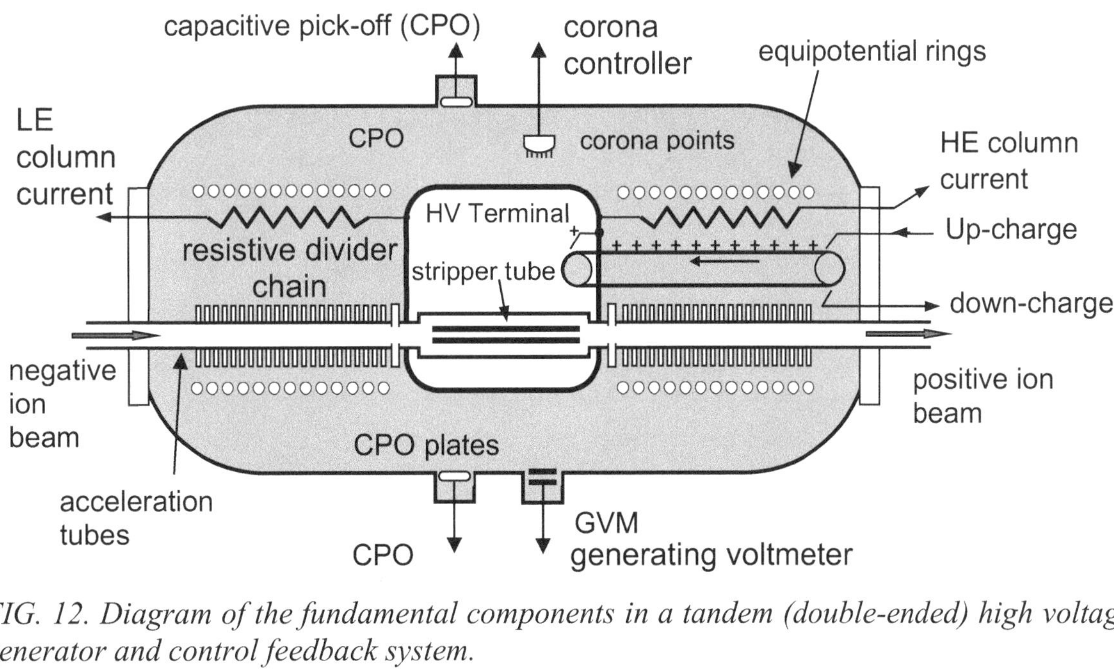

FIG. 12. Diagram of the fundamental components in a tandem (double-ended) high voltage generator and control feedback system.

ion beam energy between 0.1% and 0.001%, depending on the application. The method used is an active feedback system which employs a coronal discharge to regulate and stabilize the charge on the terminal.

Most modern charging systems use a current regulated charging power supply to maintain the charging current within 0.1% of the nominal setting. This is adequate for some applications, but a more direct method of regulation is a corona stabilization system located near the high voltage terminal. This is especially relevant to mechanical (belt or chain) charging systems. The corona mechanism is a set of sharp needle points (Fig. 13) located directly opposite the high voltage terminal which can be remotely driven toward and away from the terminal.

When the accelerator is operating, the electric field created between the high voltage terminal and the sharp needle points is highly concentrated at the tips of the points, creating an intense electric field sufficient to ionize the insulating gas molecules in the near vicinity. The field emitted electrons are quickly captured by SF_6 molecules (if using SF_6 insulating gas) or CO_2 molecules (if using a N_2/CO_2 mixture) forming negative ions that drift towards the positive high voltage terminal where they neutralize some of the terminal's charge. The transit time of the negative ions to the high voltage terminal is ~10 ms, so corona stabilization is only effective in regulating slow variations in terminal voltage.

The corona points needle assembly, if attached to the grounded tank, is a passive regulation system in that if the terminal voltage increases, the electric field at the needles will also increase, thus increasing the amount of corona current, which will in turn cancel the increased charge on the terminal. This passive feedback system is used in small accelerators where a high degree of stability and active control is not necessary. The position of the corona points needles is adjusted externally to regulate the terminal voltage to that required.

In applications where higher regulation is required, as in most analytical applications, it is necessary to include some active feedback in the corona regulator to modulate the corona current by means of an external signal. This is accomplished by including a high voltage current regulator in series with the corona needles and ground shown in Fig. 16. This regulator is typically a high voltage vacuum tube (6BK4 or 6EN8 triode) as the potential on the needles can approach 20–30 kV, which no readily available semiconductor devices can withstand. Typically, the corona current is limited to 5–10% of the total charging current, but some machines operate best closer to 15–25%.

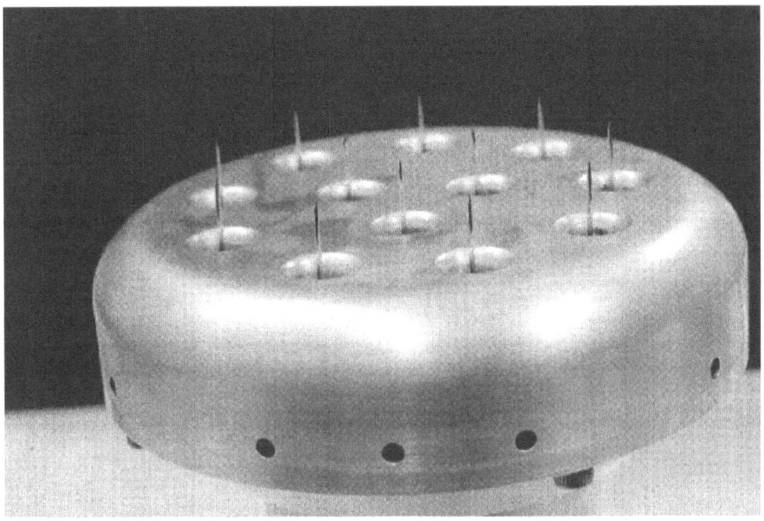

FIG. 13. Corona points needle assembly.

Modern cascade solid state charging systems, due to their high frequency of charge pumping, can stabilize the terminal voltage between 0.01% and 0.001% without the need of a corona stabilization system.

5.5.2 Corona assembly maintenance

It is important to maintain the corona needle assembly in good condition to help ensure a stable corona discharge. A poor corona discharge will typically display unusually large terminal voltage fluctuations or unstable corona currents. The causes include dull corona needles due to erosion from the discharging electric field, needles not centred in the grounded shields with sparking to the shield, high resistance in the corona circuit, low resistance between the needles and ground, breakdown across the feedthrough, contaminated or dirty metal terminal, wet or contaminated insulating gas, or defective electronics.

5.5.2.1 Checking the performance of the corona stabilization

If instabilities in the corona stabilization system cannot be identified using previous experience, then there are ways of using basic principles to locate the problems. A simple test can be done to determine whether the corona stabilization problem is related to the corona or another problem, inside or outside of the tank. The test requires comparing the performance of the corona with and without any stabilization control, as follows:

- With the accelerator switched off, disconnect the corona regulator tube.
- Ground the corona directly to the tank.
- Turn on the accelerator and bring up the terminal voltage to a low level where it is known that no corona current is being drawn.
- Observe and take note of the fluctuations in the terminal voltage.
- Extend the needles toward the terminal until the terminal potential is seen to drop slightly indicating the presence of a corona current.
- Compare the terminal voltage stability with that noted before a corona was drawn. If the variations are still abnormally large, that is, larger with corona current than without, then the problem may be inside the pressure vessel. If normal, that is, no noticeable difference, then the fault may be either the corona regulating tube, the regulation system electronics, and/or the electrical feed through on the tank.

5.5.2.2 Checking the performance of the corona feedthrough

If the results indicate a problem with the corona system and not the accelerator column components, then another test may be performed to check if the fault is in the corona electrical feedthrough. As the needles are grounded, faults with the feedthrough will not be detected in the previous test. To check for a breakdown in the feedthrough, the stability of the current drawn needs to be measured. The following test, with similar conditions to the previous test, allows an adequate measurement to be made:

- With the accelerator switched off, isolate the corona regulating tube from the corona connection.
- Place a 1000 MΩ high value resistor between the corona feedthrough terminal and the tank or ground so that the corona current will produce a potential on the needles, and the feedthrough that is typical of normal operating values.

— Repeat the test as previous. If the corona current is now unstable, then it can be concluded that there is electrical breakdown or leakage due to the high voltage, either at the feedthrough or at the needle holder insulator.

5.5.2.3 Checking the performance of the corona needles

The corona needles need to be very sharp for efficient field ionization of the gas. The tips will erode with use and typically will need to be replaced (or re-sharpened) after several thousand hours of operation. This interval is highly dependent on the average corona current. Machines that operate with a few microamperes of current can extend this interval while high current machines that operate with a hundred microamperes of corona current may need a much shorter replacement interval. When replaced, all the needles need to be adjusted to extend the same distance from the grounded shield and need to be well centred in the hole in the shield to avoid sparking to the shield.

Needles may be re-sharpened by using electrolysis. The method varies in different laboratories but essentially it requires immersing the needle points in acid while passing a high current (low voltage) through the needles to a metal plate in the solution that acts as a cathode.

5.5.2.4 Measuring corona needle resistance

During in-tank maintenance, the resistance from the needles to the external regulating tube connection (plate) needs to be checked and only needs to be several ohms. This is best checked with a low voltage multimeter, as poor connections might not be detected with a high voltage resistance meter (Megger). The electrical insulation of the corona needles from the grounded point shield, and the electrical feed through, needs to be checked with a high voltage Megger as the usual operating voltage for these parts is 5 kV–30 kV DC. There needs to be no leakage detected with 5 kV DC volts applied. Higher test voltages will breakdown in air giving irregular results, but high voltages can be used when the tank is pressurized to operating pressure with insulating gas. Any leakage detected needs to be tracked down and the defective part cleaned or replaced.

5.5.2.5 High voltage terminal dome maintenance

The stability of the high voltage can be influenced by poor conductance of the metal dome around the high voltage terminal. The metal terminal dome serves as the anode for the corona discharge and so collects the negative ions from the corona current as well as any particulates, dust, or vapours in the tank gas that acquire a negative charge from the corona points. There is usually a distinct patch on the terminal dome directly opposite the corona points. It is possible for this patch to contain oils and other materials that are effective insulators. The formation of this insulating film on the surface of the metal dome may inhibit the collection of the negative ions from the corona current. Charge may also accumulate on this insulating film until the voltage difference across the film is sufficient for the charge to penetrate the film or arc across the surface of the film to the bare metal terminal. The corona discharge may also be forced to wander, seeking out a clean area of terminal. This wandering of the corona will produce irregularities in the corona current. To maintain good accelerator performance, the terminal dome needs to be cleaned during every in-tank maintenance.

A solvent soaked rag is not sufficient for cleaning, as often the deposits are quite hard and unaffected by acetone or alcohol. The preferred method is to use an abrasive cleaning technique followed by a solvent cleaning. The abrasive normally used is fine to medium grade abrasive paper or scour pad. This can be followed with a finish (fine grade) of the same product to achieve a high polish on the terminal. This cleaning needs to be followed by wiping the dome with a rag soaked in pure ethanol or similar. It is necessary to be cautious about using other types of solvent as they may leave residues on the metal surface; the solvent to be used should first be tested on a small surface of the dome.

5.5.3 Generating voltmeters

To control the terminal voltage reliably and precisely, it is necessary to have a reference of the voltage as well as a way of regulating it. A generating voltmeter (GVM) is used, as shown in Fig. 14, which generates a real time voltage output by sampling the electric field from the high voltage terminal.

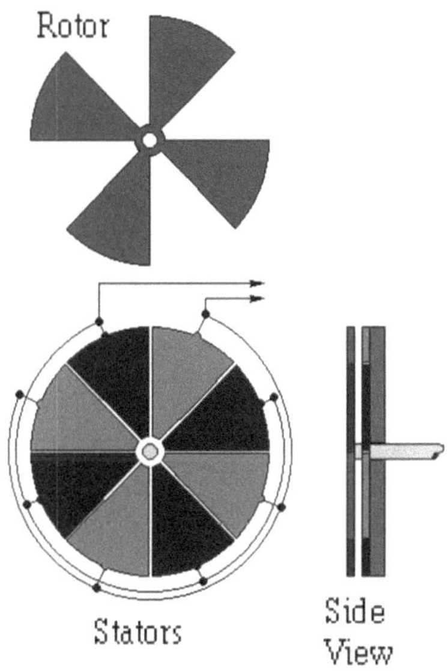

FIG. 14. Schematic of the components of a generating voltmeter.

The GVM consists of a rotating disk with vanes that look like a Maltese cross, in front of a fixed disk made up of segments as wide as a vane, with odd numbered segments electrically connected, and even numbered segments electrically connected. The GVM is located adjacent to the terminal near the inner wall of the tank. As the disk rotates, segments underneath the vanes become alternating exposed and shielded from the electric field from the high voltage terminal. Exposed segments will accumulate a charge from the electric field and shielded segments will be not. This gives rise to an alternating signal at the output of the GVM which is proportional to the sampled DC electric field present near the tank wall, and thus, proportional to the terminal voltage. This fast signal (typically 200–500 Hz) is amplified and rectified to DC, and the value is proportional to the terminal voltage. The DC voltage is displayed on a panel meter and used for comparison with a stable reference voltage.

The GVM cannot respond to voltage fluctuations of the terminal that are faster than 1–2 Hz due to the time for the average the input signal. The typical resolution of a good GVM system is 0.01%. By taking the difference between a stable reference voltage source, and the GVM voltage, an error signal is obtained which indicates the deviation of the DC terminal voltage from the desired voltage and can be used in a feedback system to regulate the terminal voltage. This feedback system is normally capable of regulating the DC terminal potential to within 0.01% on a long term (24 hour) basis.

An inherent problem with this method of voltage regulation is the susceptibility to pick up voltage fluctuations arising from vibrations of the high voltage terminal caused by the mechanical terminal charging system. The vibrations cause the base line noise to increase. To help remove the vibration noise from the GVM signal, a measurement of the vibrations is made, and the subsequent signal is then subtracted from the GVM signal. A CPO is used to make the vibration measurement.

As the electric motor for the GVM is within the gaseous environment of the accelerator, the windings dissipate their heat through the gas. When the gas is to be removed for maintenance on the accelerator, the GVM needs to be turned off, otherwise the motor may overheat and damage the windings.

5.5.4 Capacitive pick-off

The short term fluctuations of the terminal potential are regulated by feeding back an error signal, such as from a CPO, to the control grid of the vacuum tube, as shown in Fig. 15. The capacitive pick-off is typically one or two metal plates, which are electrically isolated from the tank and mounted on the inner tank wall or in a well such that their exposed surface is flush with the inner tank wall. These plates act as one of two plates in a capacitor with the terminal acting as the opposite plate and the insulating gas as the dielectric medium. If two metal plates are used, they are mounted diametrically opposite from the high voltage terminal to cancel out the influence of mechanical oscillations. By connecting a charge sensitive amplifier to the capacitive pick-off, a voltage signal is obtained that is a representation of the terminal voltage variations. Typically, the CPO preamplifier will provide a signal that is about 1% of the actual terminal fluctuations. This signal can be calibrated by applying a known signal to the terminal with the tank open and measuring the response or adjusting the preamplifier to produce a convenient signal (e.g. 1 volt peak-to-peak for every 100 volts peak-to-peak on the terminal). As the CPO is an AC responding device, it provides no information about the DC terminal potential. In practice, the frequency response of a CPO system needs to be from a few Hz to about 1 kHz. This includes the typical frequency domain of the time dependent charging system fluctuations.

A useful method to monitor, check, and/or diagnose mechanical charging system problems is to observe the output of the CPO. The sensitivity of a CPO is adequate to diagnose small discharges in the column grading resistors, and poor charging performance in mechanical charging systems. By observing the terminal voltage ripple on an oscilloscope, abnormal ripple frequencies usually can identify problems due to the pellets, or the motion of pulleys.

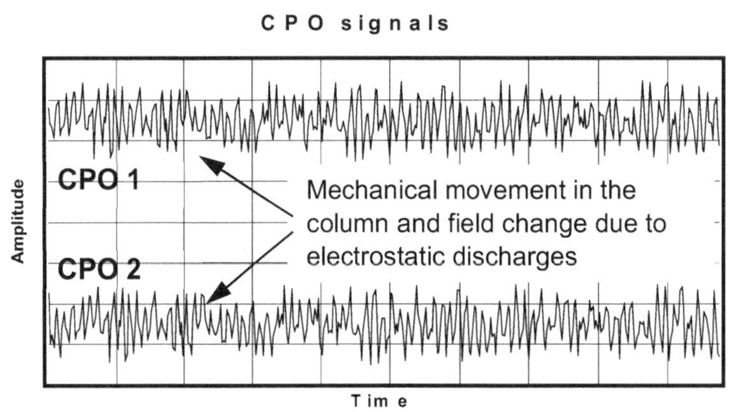

FIG. 15. A signal pattern example from a CPO for a belt charged electrostatic accelerator.

5.5.5 Terminal voltage stabilizer

A typical corona control system includes both the GVM and the CPO signals for both monitoring and control. A typical layout is shown in Fig 16.

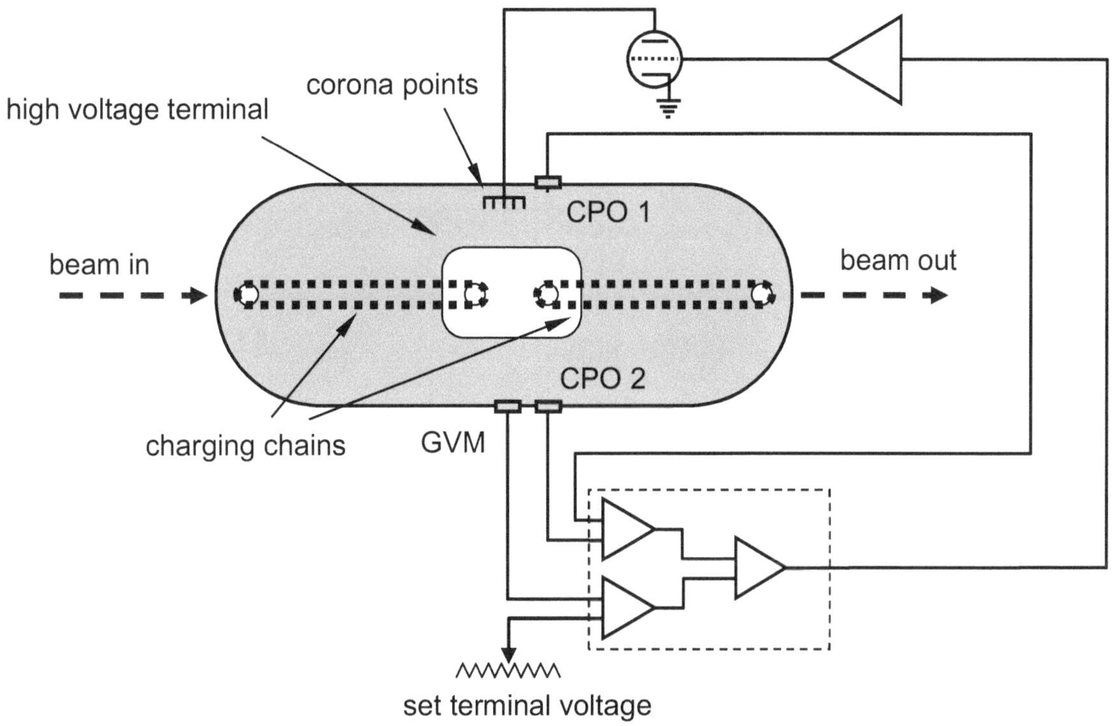

FIG. 16. Schematic layout of a typical Terminal Voltage Stabilizer unit.

The error signals are combined by means of adjustable gain potentiometers and applied to the grid of the corona regulating tube. As the vacuum tube is an inverting amplifier, an increase in the grid bias (positive voltage) will produce a decrease in the plate (or needle) potential (more negative) and thus increase the corona current flowing through the tube to the needles and the terminal. This will cause the terminal voltage to be lowered. Likewise, long term drifts in the terminal voltage can be corrected for by comparing the terminal DC voltage to a good stable DC reference. Any drift can be detected by a simple difference amplifier and applied to the corona regulating tube to adjust the corona current to compensate. Typically, the long term DC stability of this circuit can be maintained to 0.01% or better for a 24-hour period.

5.5.6 Slit stabilization

The terminal voltage stabilization system described above, and illustrated in Fig. 16, functions by sensing changes in the terminal voltage. A different system, slit stabilization, functions by sensing the change in the ion beam's trajectory when the terminal voltage changes, generating an error signal and feedback to change the high voltage to bring the ion beam back to its nominal trajectory.

Figure 17 shows the geometry of a slit stabilization system. The ion beam enters a deflecting magnet with a constant magnetic field, then follows a curved trajectory which depends on the strength of the magnetic field and properties of the ion beam. A set of slits is placed either side of the exit point which defines the nominal trajectory, with the separation of the slits set to intersect part of the ion beam. When the ion beam follows the nominal trajectory, symmetrical currents will be measured on both slits.

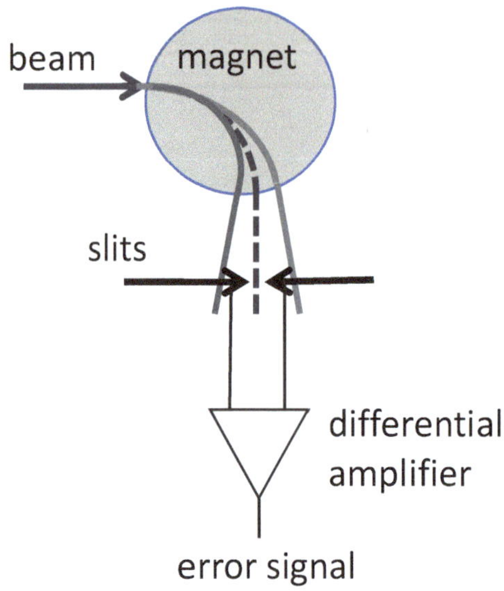

FIG. 17. Geometry of a slit stabilization system.

If the terminal voltage increases, the ion beam energy will increase, and the trajectory will change to a larger radius of curvature through the magnet and an asymmetrical current will be measured on the slits. Conversely, if the terminal voltage decreases, the ion beam energy will decrease, and the trajectory through the magnet will follow a smaller radius of curvature, creating an asymmetrical current on the slits in the opposite direction. The current on the slits are input to a difference amplifier which produces an error signal to feedback to correct the voltage on the terminal. The precision of stabilizing the terminal voltage in a well-designed and optimized system is ~0.001%.

The overall terminal stabilization system is shown in Fig. 18. Error signals from GVM and slit system are combined and used to control the power of the charging system (up-charge regulation) and/or control the discharge current of the corona regulating tube (down-charge regulation). Typically, the long term stability of the terminal voltage can be maintained to ~0.01% or better over a 24 hour period.

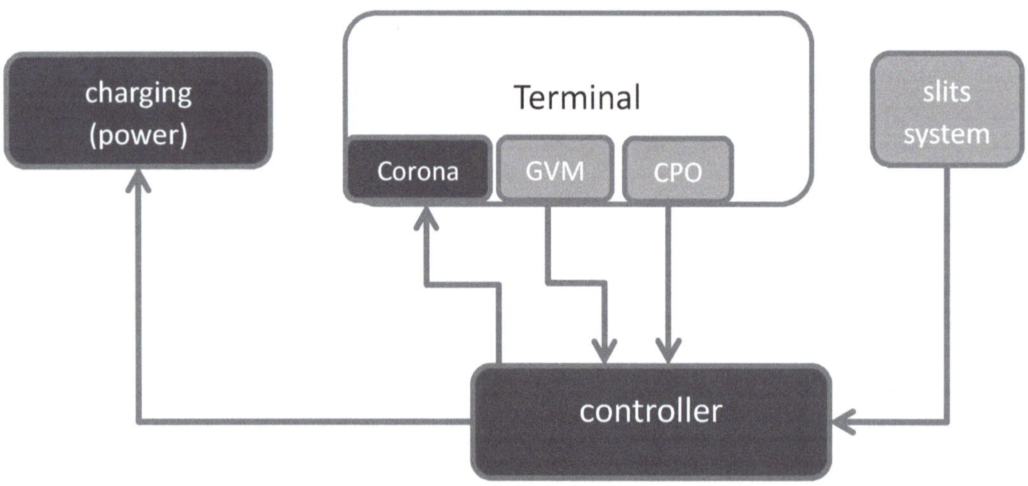

FIG. 18. Schematic of an accelerator terminal voltage stabilizer system.

6. ACCELERATION TUBES AND VOLTAGE GRADING

The acceleration tube provides an electric field to accelerate the charged particles. The basic structure of a straight field acceleration tube is shown in Fig. 19.

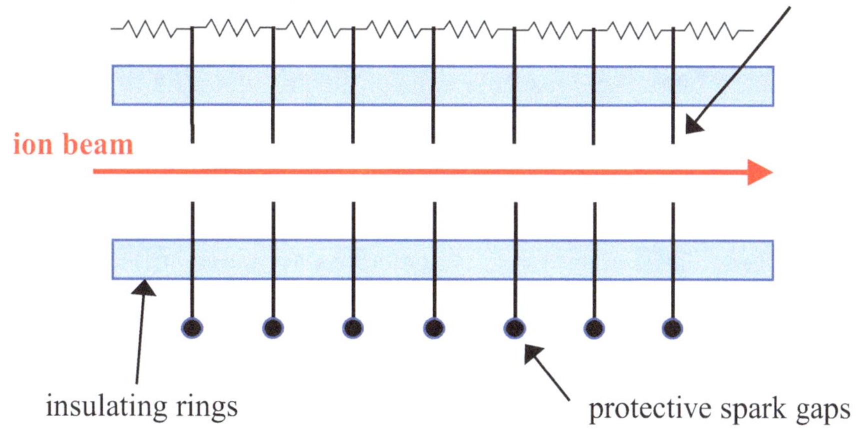

FIG. 19. Structure of a straight field acceleration tube.

The acceleration tube consists of a series of segments made up of axially symmetric thin disk or dish-shaped metal electrodes with a central hole for the ion beam to pass through and for vacuum pumping and are separated by circular insulating glass or ceramic spacers a few centimetres thick. At the ends are vacuum flanges to connect to other tubes in the accelerating structure or the accelerator terminal box, or to the ends of the accelerator pressure vessel.

Resistors are attached to the metal electrodes on the outside of the tube forming a voltage grading resistor chain that runs from the high voltage terminal to ground potential at the ends of the accelerator tank. The electrical current flowing along this resistor chain from the high voltage terminal to ground generates a potential difference between adjacent electrodes creating the electric field to accelerate the ion beam. Most tubes have protective hemispherical spark gaps on the outside between each metal electrode to protect the insulators and resistors from damage if an overvoltage condition occurs. In a transient event such as a machine spark, a large amount of energy needs to be safely dissipated, minimizing any discharges through the resistors or insulators. The spark gap will allow a flashover to occur before insulator or resistor breakdown, discharging the energy into the surrounding insulating gas.

The low energy entrance tube may also have a grid mounted inside, about three or so electrodes back from the tube entrance. This assists in effectively neutralizing the field effect at the entrance of the tube which can change the focus of the beam and adversely affect the transmission of the ion beam through the accelerator.

An inherent problem with straight tubes using flat electrodes is that the internal surface of the insulators has little protection from contamination as the gap between each segment in the tube is open. Contamination of the insulator surface can result in electrical leakages and discharges. Dish-shaped ring electrodes offer somewhat better protection from insulator contamination as the electrode partially or completely covers the gap between the insulators.

To have a long operational lifetime of a tube, it is necessary to ensure they are not operated at high voltages without prior conditioning, and that they are cleaned externally at every opportunity.

Cleaning the inside of the tubes is not easily undertaken without specialist equipment and expertise. It is important that tubes that have become unusable due to persistent breakdown are either replaced or sent back to the manufacturer for refurbishment.

6.1 ELECTRON SUPPRESSION

In the early years of accelerator development, accelerators experienced a serious problem of the terminal voltage becoming limited, even though the charging current of the high voltage generator was increased. This voltage limitation was the result of electron loading draining off current from the high voltage generator. The electron loading is due to a significant electron backstreaming current in the accelerator tube draining charge off the high voltage terminal.

When ions have collisions with residual gas in the vacuum system, or with surfaces in the accelerator tube, electrons are produced. These electrons are accelerated by the electric field in the accelerating tube to high energies and as they collide with surfaces in the accelerator tube, they produce secondary electrons creating a multiplying effect. Also, the electron collisions with surfaces produce high energy X ray bremsstrahlung which creates some ionization of the insulating gas causing some additional current to be drained off the high voltage terminal. The X ray bremsstrahlung can be intense posing a significant radiation hazard to personnel. Consequently, all modern acceleration tubes are designed with electron suppression systems.

Effective suppression is achieved by having a high vacuum pressure to minimize ion-residual gas collisions, minimizing secondary electron production using materials with a low secondary electron production, but most importantly, by deflecting and removing electrons produced inside the tube before they can gain sufficient energy to generate a multiplying effect. Deflection techniques use either weak magnetic fields installed within the tube or by using inclined field tubes.

6.2 LINEAR INCLINED FIELD TUBES

As mentioned above, straight tubes allow significant electron acceleration and currents along the tube. Linear inclined field tubes, invented in 1961 [19], have electrodes that are inclined relative to the to the tube's centre axis, as shown in Fig. 20.

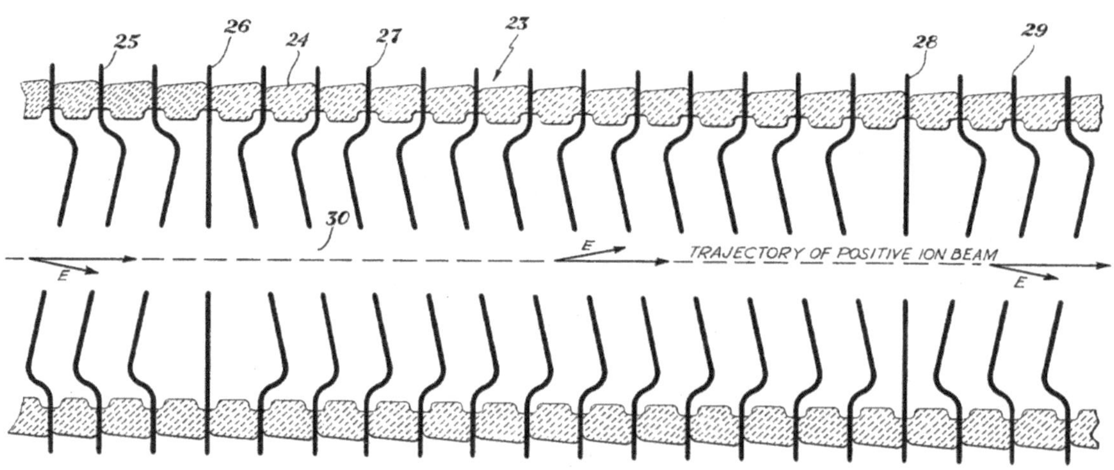

FIG. 20. Linear inclined field tubes [19]

The inclination of the electrodes creates a small transverse electric field E (indicated in Fig. 20), which is sufficient to deflect and remove electrons onto electrodes before they can acquire sufficient energies to be of concern. This small transverse electric field also affects the ion beam, steering it slightly off axis. To compensate for this steering, the tubes have groups of electrodes with an inclination in one direction, followed by groups of electrodes inclined in the other direction so there is no net deflection. The electrode inclination angle is 12°.

In modern designs, magnetic electron suppression can also be used for electron suppression. Small permanent magnets are located on the accelerator tube to create a weak magnetic field that sweeps electrons out of the beam axis and into the electrodes. As is the case in electrostatic suppression, the following tube section has the direction of magnetic field reversed so that there is no net deflection of the ion beam.

6.3 SPIRAL INCLINED FIELD TUBES

Spiral inclined field tubes are like linear inclined field tubes except that each electrode is rotated with respect to the previous one. This makes the direction of the small transverse electric field E (indicated in Fig. 20) rotate about the beam axis forming a geometric shape of helical spiral. After a section of electrodes have completed a rotation of 360° the ion beam is slightly displaced off-axis. This displacement is corrected for by using a second section spiralling in the opposite direction. After the second section completes a rotation of 360° the beam's net displacement and angular deviation is negligibly small.

Spiral inclined beam tubes have demonstrated superior transmission over other types of acceleration tube, especially at low energies and for heavy ions. These tubes have demonstrated superior beam optics and residual gas pumping due to their configuration and have a proven record of stable operation, showing no signs of coupling to charging current irregularities. Magnetic suppression can also be used as in other inclined field tubes.

6.4 ACCELERATION TUBE MATERIALS

The type of materials used to construct acceleration tubes are required to have certain properties such as an ability to be machined and formed, good compatibility in high vacuums, low secondary electron emission coefficient, etc.

6.4.1 Electrodes

Three metals are mainly used for the construction of electrodes: aluminium, stainless steel and tungsten. Aluminium, being a soft metal with a relatively low melting point, is prone to sputtering from arc damage which deteriorates the electrodes and leaves evaporated layers on insulator surfaces which are difficult to remove. Stainless steel has excellent high vacuum compatibility and is more commonly used for vacuum flanging at the ends of the acceleration tubes and for spark gaps. Titanium has a high melting point and does not sputter like aluminium. The maximum secondary electron emission yields for these technical grade materials are 3.4 @ 350 eV for aluminium, 2.0 @ 300 eV for stainless steel, and 2.0 @ 250 eV for titanium [20]. The low secondary electron yield of titanium, together with its other properties such as high melting temperature, low sputter yield, and compatibility with bonding agents, make titanium the preferred choice for electrodes.

6.4.2 Insulators

By far the most common type of insulating material used in accelerators is borosilicate glass. It is easy to manufacture, mechanically strong and has a dielectric strength ~10 kV/mm. A limitation of the glass and metal construction of the tube is that the insulators and electrodes are cemented together with PVA (polyvinyl acetate) resin. The vacuum surfaces cannot be cleaned by baking, and other cleaning methods can result in residues leaving paths for discharge tracks at higher field strengths. Tubes that are used at high terminal voltages require some conditioning to reduce electron load and bremsstrahlung levels.

Ceramic tubes are made from alumina-ceramic (Al_2O_3) which has a similar dielectric strength to glass (values can vary depending on the grade of material). Ceramic tubes consist of titanium metal electrodes with aluminium-bonded joints between them and ceramic insulating rings. The length of a basic module is ~20 cm. Three modules joined together constitute a standard 1 MV column module. An advantage of ceramic tubes over glass tubes is that the ceramic tubes are bakeable, enabling them to be heated for cleaning.

6.5 CONDITIONING

Conditioning is a procedure to increase the electric breakdown voltage and voltage stability in the accelerator system. It is always used when the accelerator is returned to service after maintenance, or when the terminal voltage is to be increased to a level substantially higher than that currently being used.

Conditioning is a cleaning procedure for the physical removal of high voltage breakdown sites on insulators, and removal of surface contamination and desorption of surface gases by high voltage induced stress. As the terminal voltage is increased, small micro-discharges occur across insulating surfaces producing local ionization and a cleaning action. This can be indicated to the operator by a transient increase in vacuum pressure and a transient burst of bremsstrahlung X rays. Other effects may also be observed including changes to the up-charge current and surges in the signal from the CPO from small discharges. The discharges, if sufficiently intense, have the potential to leave a deposit across the insulator surface creating a tracking path for high voltage breakdown. Thus, the high voltage needs to be incremented slowly, in small steps, and with pauses between each increment. Even when there is little or no apparent change in the vacuum or radiation levels during the conditioning process, some micro-conditioning may still be occurring. Care needs to be taken during conditioning with pauses between voltage increases after stability is observed, and not to expedite the process as irreparable damage to the tube may occur.

6.5.1 Conditioning procedure

This is a general guide only to conditioning, as each accelerator will have its own unique conditioning levels and experienced operators may have their own well-established techniques.

For effective monitoring of the conditioning, a sensitive vacuum gauge is required with the gauge head located near the tube(s) exit on the beam lines adjoining the accelerator tank, and a radiation monitor with the detector head needs to be located at tube height on the outside wall of the tank, near the high voltage terminal.

The conditioning voltage needs to be at least 100 kV higher than the required operating voltage. It is important to ensure the conditioning voltage does not exceed the design and operational limits of the accelerator. Conditioning needs to be made without the beam in operation.

It needs to be ensured that the vacuum gauge, radiation monitor, oscilloscope showing the CPO signal and the charge current meters are switched on and can be easily seen from the control console. The radiation, vacuum, and CPO signals are the main references used during conditioning, which involves the following steps:

- Start with a low voltage and work upwards in small steps;
- Reduce the voltage level immediately if discharges are heard or observed;
- Allow sufficient time for conditioning levels to return to a low and steady state;
- If conditioning levels persist, it is likely that there is a fault or damage somewhere in the accelerating tubes or columns.

The suggested conditioning procedure is:

- Turn on the accelerator power supply charging system and increase the terminal voltage current until about 20% of the maximum accelerator terminal voltage is reached. If this voltage cannot be reached without abnormal levels appearing on the monitoring devices, reduce the voltage until background radiation levels, normal operating high vacuum levels, and a low noise CPO signal is reached. These levels need to be recorded as the baseline.
- Increase the voltage by a few percent of the maximum rated voltage of the accelerator then wait and observe the monitoring devices for changes. There will be some changes to the conditioning levels which are normal for increased voltages and are safe, but larger transient levels or periods of extended high levels are potentially damaging, and the voltage needs to be immediately reduced. The conditioning levels need to be allowed to return to a near quiescent state before any further voltage increases. There are cases where damaged tubes exhibit persistent radiation which only reduces after a large voltage reduction. If the radiation level does not reduce, or fluctuates during the conditioning procedure, the tube may be damaged and faulty segments may have to be shorted out.
  - The vacuum needs to be kept within the 10^{-4} Pa range or better. Some excursions beyond this range may occur intermittently. The vacuum needs to be allowed to return to a level close to the baseline level recorded at the start before further voltage increases. There may be a small increase in the baseline vacuum as the voltage increases.
  - The CPO signal may indicate a large amount of noise but if high peaks remain or become regular, the voltage needs to be lowered as internal discharging may be occurring. The up-charge current can be allowed to surge but only if it returns to the normal level after each surge and the frequency of surges become significantly less after each excursion, otherwise the voltage needs to be immediately reduced.
  - If tank sparks or other discharges are heard, then the voltage needs to be decreased by at least one step value until the discharging stops.
- When all the conditioning levels have decreased to acceptably low and 'normal' levels, repeat the second step. Be aware that as the voltage increases the frequency and surge levels of conditioning may increase. Because of this, the voltage increments need to decrease to maintain conditioning levels within the benchmark levels. Continue the conditioning until a voltage at least 100 kV above the required operating voltage is reached.

6.6 HANDLING AND MAINTENANCE OF ACCELERATION TUBES

Accelerator tubes comprising glass or ceramic bonded to metal are susceptible to de-bonding under relatively light lateral loads due to insufficient support, pressure changes in the tube and/or high humidity environments. The forces experienced when changing from vacuum to positive pressure in a tube can cause some bonds to be overly stressed causing fracture. If acceleration tubes need to be raised to atmospheric pressure, then dry nitrogen or argon needs to be backfilled to ensure a low humidity and oil and contamination free atmosphere inside the tube. The gas needs to be regulated and introduced very slowly to minimize potential pressure shocks to the bonds.

Ideally, all movement of acceleration tubes needs to be done with the tube resting on a support frame which evenly supports the whole tube however, when this is not possible, the tube needs to be supported at least every 500–750 mm. If returning a tube for refurbishment that is leak free, then the tube needs to have blank flanges on the ends and a clean, dry, inert atmosphere inside.

6.6.1 Cleaning acceleration tubes

The use of contaminated or dirty tubes may cause instabilities in terminal voltages and hence unstable ion beams and increased secondary electron emission. Tubes then need to be cleaned or replaced. One method of cleaning tubes is by heating them to a few hundred degrees Celsius while under vacuum to remove carbon and hydrocarbon buildup; this may be achievable for ceramic tubes but not glass. Glass tubes are bonded together with adhesives that cannot withstand high temperatures or immersion into liquids including solvents. During accelerator maintenance, the outside of the tubes may be cleaned with ethyl alcohol and wiped clean then blown dry with air as to ensure all residual solvent is removed. The inside of the tube may be contaminated with layers of carbon and hydrocarbons and some sputtered metal, especially when used in vacuum systems that have poor vacuum or use plain ring type electrodes. The inside of the tube is difficult to clean even when the ends can be reached.

The dark brownish radiation damage in the glass due to bremsstrahlung radiation does not typically render the tube as unusable unless discharging is also occurring and causing large instabilities in the ion beam energy. If the glass tubes are so heavily contaminated that cleaning is essential, it may be more beneficial and cost-effective for the tube to be sent to the tube manufacturer for refurbishment.

If damage has occurred to the tube insulators, such as tracking marks left by discharges or spots left by arcing, then unless the damage can be physically reached and polished away with a fine abrasive, those effected insulators need to be shorted out across the spark gaps.

6.6.2 Finding faulty insulators on the accelerator tube

Generally, an accelerator tube with damage to insulators can cause high stress points in the electric field which can generate a high intensity X ray radiation field outside of the accelerator. Using an X ray detector, a map of the radiation level is made along the accelerator tank at tube height to locate the approximate position of the affected tube. During in-tank maintenance, glass tubes can be visually inspected near the located position for tracking marks on both the inside and outside of the glass insulators. Some damage might not be visually detectable however, those segments that are, need to be electrically shorted together by placing a metal sleeve across the electrode spark gaps of the effected insulators.

6.6.3 Vacuum leak testing of acceleration tubes

When a new accelerator tube is to be installed or a vacuum leak is suspected in already installed tubes, a thorough vacuum leak check needs to be carried out. Vacuum leaks in tubes can be difficult to locate especially in those that are already installed. Stresses caused by the high pressure insulating gas can cause leaks to appear that cannot be easily located at atmospheric pressure during in-tank maintenance. For this reason, care needs to be taken to thoroughly leak check all sections and joints.

New tubes need to be leak checked before installation to avoid unnecessary effort to remove it later if a leak is detected. A quick method is to first seal one end of the tube with a blank flange and the other with a pump-out tube that can be connected to a helium leak detector, then cover the whole accelerator tube with an envelope of plastic that can be sealed and filled with helium. Immersing the whole tube in a helium atmosphere quickly checks the whole tube at one time. If a leak is found, then traditional methods of leak finding can be pursued. If a leak is found and cannot be repaired, then the tube needs to be returned to the manufacturer. If all tubes and other components are leaktight, then the tubes can be installed. Once installed, the whole assembly needs to be leak checked again.

During the pressurization of the tank, the vacuum in the beam lines adjoining the tank needs to be monitored for a rise in vacuum pressure which is an indication of a pressure dependent leak. If a leak is observed, then the insulating gas needs to be removed and the tube assembly thoroughly leak checked again. If the leak is still not apparent, then the tubes need to be removed one-by-one or isolated in-situ and individually checked. The risk of causing damage to tubes increases the more they are handled.

6.7 RESISTOR DIVIDER CHAINS

A robust resistor voltage divider is an essential element for the effective and stable operation of an electrostatic accelerator. The resistor divider establishes a uniform voltage gradient along the length of the accelerator creating a potential difference between acceleration tube electrodes to accelerate the charged particle beam.

Resistor assemblies used in the hostile environment inside an accelerator pressure vessel need to possess certain qualities:

— Resistance values that correspond with the accelerator charging configuration;
— Resistance values that remain stable over a long period of time;
— Resistor values that have a high degree of precision;
— A mechanical arrangement that eliminates the possibility of dry contacts;
— An ability to withstand high voltage discharges along the accelerator column;
— A physical robustness in the mounting configuration.

Resistor failure can result in poor electron suppression, poor beam focusing, and high voltage discharges, which can cause damage to the insulators in the accelerating tubes or columns. If faults are left to persist, microcracks can develop in the insulators due to numerous tracking discharges.

Older style resistors are typically of the carbon layer type, which become unstable over time. These have now been superseded with metal-oxide film resistors (typically rhenium oxide), which are very stable, fabricated with high precision, and have excellent voltage holding characteristics.

Changes to the design of resistor assemblies have removed the possibility of dry contacts which occur when the resistor relies on spring tension to contact the supporting structure. Dry contacts can lead to increased sparking with subsequent problems. This phenomenon is particularly apparent when the resistors are used in an insulating medium of SF_6 gas. Ideally, all contacts need to be screwed connections.

Other designs use a coaxially shielded assembly with a circumferential spark gap as shown in Fig. 21. In this design, a metal tube is connected to one end of the resistor and coaxially encloses the resistor over a large part of its length. A disc mounted at the other end of the resistor forms a spark gap with the tube. This arrangement provides an alternate path for high frequency, high voltage discharges rather than through the resistor.

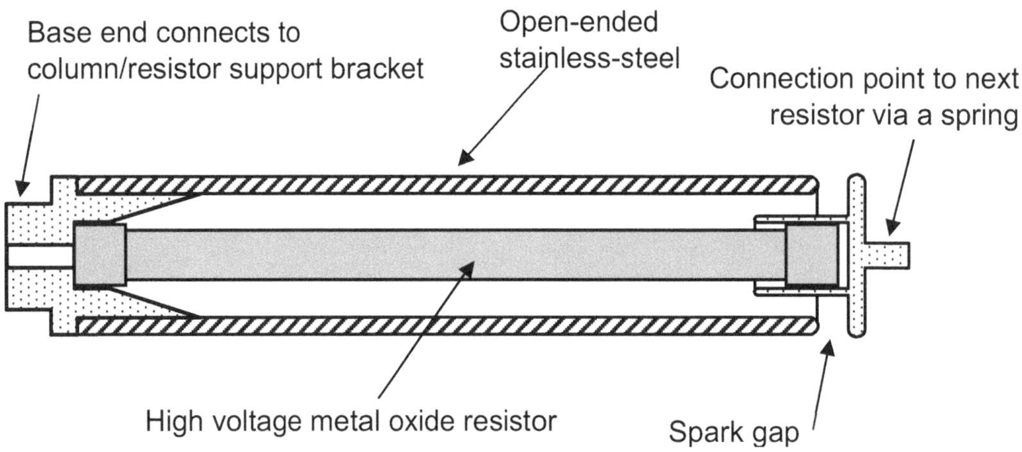

FIG. 21. Coaxially shielded resistor assembly with a circumferential spark gap.

6.7.1 Testing of resistors in-situ

Resistance values that are higher than the nominal value generally indicate a damaged resistor due to spark damage on the resistor's metal oxide film. Continued use will increase the damage causing the voltage to become unstable and cause the beam quality to degrade.

Resistors that are lower in resistance than the nominal value generally are either very old (the resistance generally declines with age) or are contaminated with a conductive film. It is also possible that the column or tube is contaminated with a film or debris that is causing excessive leakage across a parallel path.

When undertaking maintenance on the resistor chain, it can be useful to locate defective resistors using two different voltages. Readings taken at low voltage (<50 V) will quickly locate resistors that have microscopic breaks in the film, as this test voltage is insufficient to bridge any gaps in the film. Resistance readings using a high voltage (1–5 kV) can locate problems in the column or tube that may be causing sparking, corona, or leakage at operating potentials. Typical problems found are foreign material in the column (including lost screws or small tools inadvertently left behind), loose spark gaps, or damaged tube insulators. Testing at high voltage is considered essential.

Resistor values used in most accelerators, and particularly those where limited up-charge is available, are typically 500–1000 MΩ. It is important to have an instrument capable of measuring to an accuracy of about 1% of this high resistance value. Meggers/insulator testers with a test voltage of 1–5 kV are typically used. Alternately, a high voltage power supply and a series microammeter can be used. For low voltage readings, there is a requirement to accurately measure currents less than 100 nA. This low current can provide a difficulty if a suitable Megger is not available.

Resistor values need to be recorded and logged so that changes over time can be tracked and potential problems identified early. Ideally, resistance values are measured and recorded just before the accelerator pressure vessel is closed after maintenance. This has an advantage of checking that all column and tube components are reassembled correctly and that no foreign objects have been inadvertently left behind. Resistor measurements need to be made at least annually.

6.7.2 Preventive maintenance of resistor dividers

The following maintenance needs to be carried out whenever there is a tank entry:

- Uncoated resistor elements need to be cleaned of any surface contamination;
- Alignment of resistor assemblies and spark gaps needs to be checked;
- Screw connections where used need to be checked for tightness;
- Resistor elements that are connected by spring-loaded sockets/pegs need to be checked for tightness, and pegs and sockets checked for any signs of sparking;
- Check spark gaps for evidence of abnormal sparking and if apparent, the resistor assembly needs to be thoroughly inspected and replaced if necessary.

7. TANDEM ACCELERATOR ION STRIPPING SYSTEMS

The high voltage terminal inside a tandem accelerator is at a positive voltage with respect to ground. Negative ions injected into the tandem accelerator are attracted towards this positive high voltage terminal. Within the terminal is a region where electrons are removed (stripped) from the incident ions, converting them into positively charged ions. These positively charged ions are then repelled by the positively charged terminal and exit from the other end of the accelerator tank. Ions thus gain energy in two steps: first, from the attraction towards the high voltage terminal and second, after the electron stripping process, from the repulsion away from the high voltage terminal. The electron stripping process produces ions with a variety of discrete charge states and hence, the exit beam from the tandem accelerator contains ions with a variety of discrete energies. Electron stripping is done by one of two methods: either by passing the negative ions through a thin carbon foil, or by passing them through a tube of gas.

7.1 CARBON FOIL STRIPPING

As the negative ion beam passes through the carbon foil, the loosely bound electrons in the ion's outer atomic shells are removed producing positively charged ions. Typically, a self-supporting thin carbon foil of thickness ~5 $\mu g/cm^2$ is used. These foils are mounted on vertical frames and contained inside a cassette within the high voltage terminal. When a foil breaks it can be remotely removed out of service and replaced by a new foil from within the cassette.

Thin carbon foils can be purchased commercially. Previous experiences have shown that it is generally more cost-effective, and quality-effective, to procure foils rather than attempt to manufacture them. Procured foils are usually sold mounted on glass microscope slides. These foils need to be removed from the glass slides and mounted on foil holder frames. To ensure a high success rate in their removal, it is important to understand how the fragile foils are mounted on and removed from the glass slides.

Between the carbon foil and the glass slide is a thin film of a water-soluble material. Immersion of the glass slide in water will separate the carbon foil, leaving it free to float on the surface of the water. The foil can then be mounted on its supporting frame as described below. To add mechanical strength to the foil during handling, the foil has been coated with a sacrificial material which is removed by heating at high temperature. This heating is the final step of the foil mounting process.

The process used to mount the carbon foil onto their holders varies between accelerator facilities, but essentially the technique is the same, as follows:

- First, cut the foils to size by scratching, or scoring the foil membrane while still mounted on the glass slide. The size of the foil needs to be large enough to cover the whole area of the foil holder.
- Second, very slowly immerse the glass slide into water to allow the foil to peel away without it breaking, and leave it floating on the surface of the water.
- Third, carefully slide the foil holder under the floating membrane, lifting it slowly to capture it over the centre of the supporting area. If the captured foil is not properly affixed, then refloat the foil and repeat the capturing process.
- Fourth, when it is captured correctly on the foil holder, allow the excess water to drip away then stand the foil holder upright in a supporting stand, allowing it to dry.

— Finally, remove the sacrificial coating by placing the dried foils into a cold oven and heat to 400°C for 2 hours. Using an initially cold oven minimizes thermal shock to the foil and helps prevent possible breakages.

The foil mounting procedure is best done in an environment of still air, such as in an open glove box. Any draughts of air, such as from air-conditioning systems or from persons breathing on the foils, can easily break them. Mounted foils are very fragile and are best transported inside a closed box.

7.2 GAS STRIPPING

Figure 22 shows a typical gas stripper system. The gas stripper is a long, open-ended tube (stripper canal) into which a gas such as argon or nitrogen flows at a low pressure. The gas flowing out from the ends of the stripper canal can enter the accelerator tubes causing an extra gas load on the main accelerator vacuum system. To circumvent this, a turbomolecular pump is sometimes used to collect gas exiting from the stripper tube and recirculate it back into the gas input delivery tube.

If the density of gas molecules inside the stripper tube is sufficiently high, there is a high probability of collisions between the incident ions and gas molecules. In the collision process, electrons are removed from the negative ions producing positive ions with different charge states. These positive ions are then accelerated in the high energy accelerator tubes.

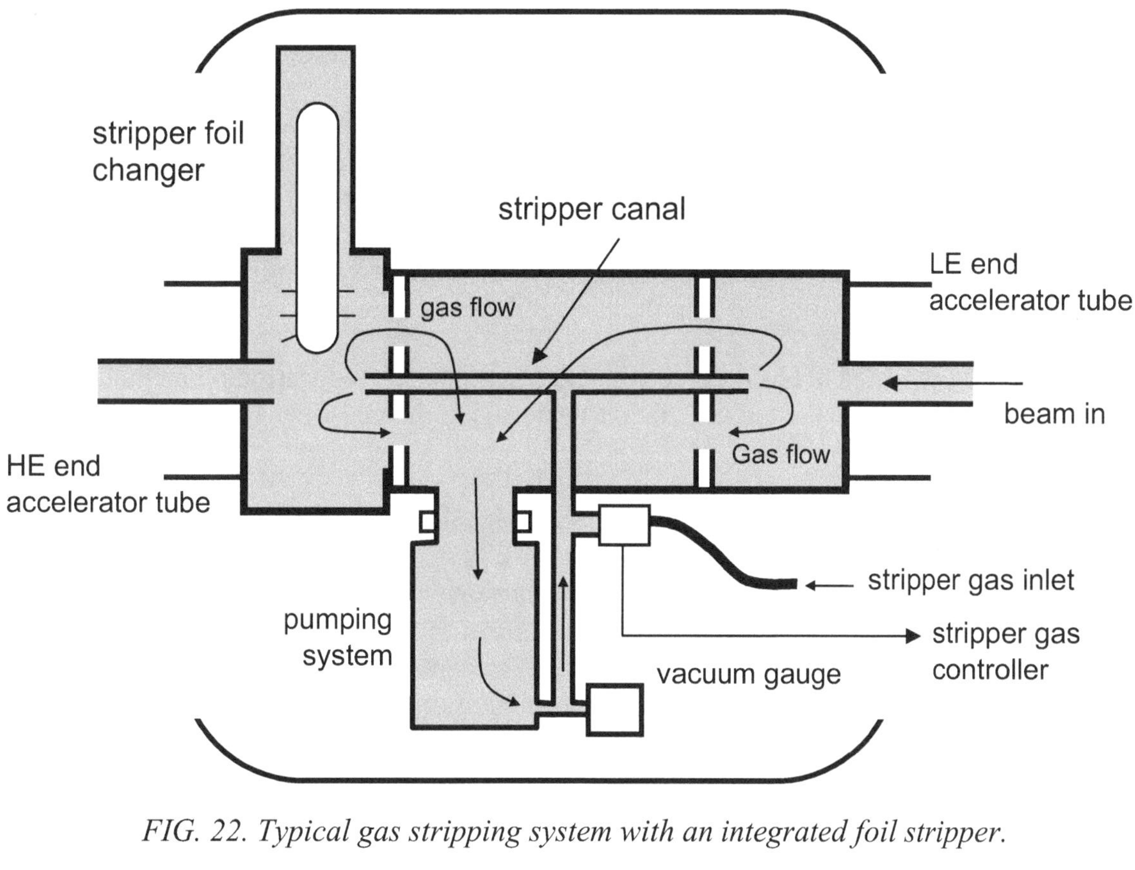

FIG. 22. Typical gas stripping system with an integrated foil stripper.

Two main factors determine the charge state distribution of the positive ion beam. They are the incident energy of the negative ion beam, and the effective thickness of the stripping medium. The higher the incident ion energy, the higher the average value of the charge state distribution, i.e. the higher the incident ion energy, the greater the number of electrons stripped off the incident ion. If the density of the stripper gas is too low, there are too few collisions, and the output yield is low. Too high a gas density can cause electron reattachment to already stripped ions and the average value of the charge state distribution decreases. Additionally, the relatively high number of collisions can cause significant energy and angular straggling of the ion beam degrading its quality. The optimum value of gas stripper density is found by experiment. It will also vary according to the type of ion being stripped.

7.3 FAULT FINDING IN STRIPPERS

Faults in the accelerator's stripper usually mean there is no stripping medium present: no foil or a broken foil for foil stripping, and no gas or insufficient gas or a gas stripper.

7.3.1 Foil stripping

To determine the effectiveness of the foil stripper, it is necessary to inject a negative ion beam into the accelerator and measure the output beam current in the high energy Faraday cup. If all accelerator settings are normal and no beam current is measured in the cup, the foil may have completely ruptured. If some current is measured with normal settings, the foil may only be partly damaged. In either case, changing to the next foil in the cassette will generally confirm a poor foil, unless of course it too is also broken.

To assess the foil changer for correct operation, activate the changing mechanism whilst observing the current in the high energy Faraday cup. There will be a momentary reduction in current as the foil frame passes through the beam. If there is no reduction in current, and there is no reason why the changer is not working, it is then necessary to view the foil through the centre of the accelerator tubes. It is important for both foil and gas strippers, that the accelerator is fitted with viewing ports at each end of the machine on the straight through beam line extensions. With a diffused light shining at one end of the straight through beam line extension, it needs to be viewed through the other end with a telescope while actuating the foil changer. Intact foils will be seen as a darkened image as they pass through the beam of light. If it is not possible to have permanent viewing windows, then the low energy and high energy sections of the straight through beam lines will have to be removed.

If the foil changer is suspected of being defective, it is essential that the location of the fault is isolated, to either the mechanism inside the tank or that part external to the tank. The identification of internal faults will involve a tank entry.

The first test is to check whether the external actuating signal that operates the foil changing system inside the tank is present. The checking point is typically at an electrical feedthrough into the accelerator tank. Sometimes, a feedback signal is sent from the tank, usually at the base end, to indicate that at least some part of the internal mechanism is operating correctly. If this signal is present, then the fault is most likely at the terminal and tank entry is required. It is important that this type of feedback be fitted to confirm the correct operation of the internal foil changer mechanism.

7.3.2 Gas stripping

To determine the effectiveness of gas stripping, it is first necessary to determine that gas is flowing inside the stripping tube. The usual method is to activate the gas supply and if working, a pressure rise at both the low and high energy beam line vacuum systems will be observed. In some accelerator systems, gas pressure is measured inside the stripping box and an indicator signal is sent to the outside of the tank via an optical fibre link, or light pipe.

When the stripper gas control is activated and minimal or no change is seen in the external vacuum system, it can be assumed that:

- The external gas control system is at fault;
- The internal (inside the tank) gas control system is at fault;
- The internal gas control valve is faulty; and/or
- The gas supply is depleted.

The fault can be isolated to an internal, or external fault, by determining that the actuating signal that operates the system inside the tank is present at the last point of tank entry, typically at an electrical feedthrough.

On accelerators where the stripper gas bottle is located externally, the gas supply is delivered to the stripper canal via a plastic tube run along the low energy accelerator tubes. Excessive stripper gas consumption is a symptom of a leakage in the plastic gas delivery tube. Tank entry is necessary to check if the tube is damaged or defective.

When the stripper gas control is activated, and an abnormally large pressure change is seen in the external vacuum system, it is most likely that the internal gas control valve is faulty. Alternatively, if a recirculating gas stripping system is in use, it is possible that the turbomolecular pumping system is not operating. The procedure is then as follows:

- With the stripper gas flowing, inject a negative ion beam into the accelerator and measure the current in the high energy Faraday cup.
- If all accelerator settings are normal, and poor beam current is measured in the cup, change the gas flow settings, and monitor the effect.
- If there is still no current measured, but gas is flowing, the fault may be a misalignment in the gas stripper tube, or the beam is not being transported along the correct path on which the stripper is aligned. It may be possible to observe a small off-axis alignment in the stripper tube position by viewing through the accelerator using a telescope as described previously. Caution needs to be exercised in this decision-making process, as the collimating aperture in the stripper box may obscure a gross misalignment of the tube.
- If the tube is aligned, then the beam alignment needs to be checked and if necessary, readjusted and the test repeated.

8. INSULATING GASES

For accelerators to be charged to, and reliably maintain the high voltages necessary for operation, all high voltage components need to be sufficiently electrically isolated from ground. To achieve this, the accelerator terminal, support column and tubes are located within a pressure vessel that is filled to a high pressure with a gas having high electrical insulating properties. The insulating gas pressure and diameter of the pressure tank both need to be sufficiently large to create a high electrical resistance between all potential electrical discharge pathways.

As the maximum terminal voltage of the accelerator increases, the required diameter of the pressure vessel — and concomitantly, the required volume of insulating gas— increases. Some accelerators have an asymmetrical tank shape to reduce the volume of insulating gas required. They have a large diameter at their centre region where voltages are highest, and a smaller diameter near the ends where voltages are lower. Although this design involves a smaller volume of insulating gas compared to a symmetrical tank design, a symmetrical tank can be easier to manufacture and have advantages in terms of a larger in-tank working space for personnel.

Only a few suitable insulating gases are available. These gases are typically the same as used in the electrical power industry as insulators of high voltages in industrial plant and electricity supply equipment. One of the main properties possessed by an insulating gas is its high dielectric strength, or electronegativity. In the early years of accelerator development, it was common to use mixtures of N_2 and CO_2 as the insulating medium. Nowadays, most modern accelerators use sulphur hexafluoride (SF_6) as the insulating gas. The superior electrical insulating and arc quenching properties of SF_6 have made it the preferred insulating gas. Its better insulating properties have enabled smaller, and so less costly, pressure vessels to be utilized.

The insulating gas needs to be kept dry, and free of contaminants, such as hydrocarbons, to maintain its high electrical insulating properties. Its moisture content, as measured by its dew point, needs to less than -45°C. Dew points between -50°C and -60°C are typically required before routine high voltage accelerator operations can commence. It is not appropriate to operate accelerators with a dew point higher than -40°C.

Before any maintenance work is carried out (either involving entry into the accelerator tank or involving the insulating gases), it is essential that personnel are aware of the potential safety hazards and precautions to use. There have been incidents in which personnel have been overcome by the toxic by-products created by electrical discharges in the insulating gas, or by asphyxiation caused by inadequate venting and flushing of the accelerator tank with air before entering. It is essential that the recommendations for the safe handling of insulating gases provided by the gas manufacturers and competent authorities be implemented as part of standard laboratory practice, as well as working in confined spaces.

8.1 NITROGEN/CARBON DIOXIDE

Nitrogen and carbon dioxide gas mixes are typically a mix of 75% N_2 and 25% CO_2 by volume. The advantages of these gases are that they are readily available, inexpensive, relatively safe to handle, and do not create any toxic by-products in the event of electrical discharges. These gases are generally supplied in individual pressurized gas cylinders. Plant is available for producing nitrogen gas from bulk, such as utilizing the evaporation in liquid nitrogen storage tanks.

When tank entry is required for maintenance, it is common practice to vent these gases to air, and later refill the accelerator pressure vessel with a new load of gas rather than recycle the gases.

Recovery plants are available to recycle the gases, but it is generally not a cost-effective option. Nitrogen and carbon dioxide are supplied in various grades (purities) depending on their intended use, such as industrial, medical, or food processing. As there can be variability in the quality of the gases provided by the vendor, it is important that a dew point moisture content measurement be made after each gas fill. If the dew point is too high, it may be necessary to vent some gas from the pressure vessel and refill with new gas. This procedure may need to be repeated until a dew point lower than -45°C is achieved. If the facility is equipped with a gas recovery plant, recirculation of the gas mixture through a dehumidifier system can replace venting of the gases to atmosphere.

8.2 SULPHUR HEXAFLUORIDE

The main hazard associated with the use of SF_6 is the toxic fluorine breakdown products that are created during electrical discharges inside the accelerator pressure vessel. If they are not fully removed from the accelerator tank before personnel enter, potentially serious injuries can occur. These by-products are also corrosive and if left to accumulate over a long period, can corrode components causing much time and money to affect repairs inside the accelerator and the gas recovery system.

Due to the high cost of SF_6, it is recycled during maintenance operations. The gas handling plant needs to have a method of drying and purifying the gas to remove moisture and the fluorine breakdown products. The commonly used purification system is to circulate the gas through a bed of activated alumina. After many uses, the alumina bed becomes saturated and needs to be regenerated. During the regeneration process, care needs to be taken to ensure that the volatile products released are safely vented outdoors. The alumina needs to be periodically replaced to maintain a high efficiency in the gas purification system.

8.3 OTHER GAS MIXTURES

A small percentage of SF_6 can be added to N_2 and CO_2 gas mixtures to enhance dielectric strength. Insulating properties comparable to 100% SF_6 can be obtained. The maximum terminal voltage can typically be increased by up to ~25% using this three-component hybrid mixture. The accelerator filling pressure of this three-component gas mixture needs to be the same value as the two-component mixture, with the dew point kept less than -45°C. The normal N_2 and CO_2 gas handling system can operate successfully with this three-component mixture, as its density is not much greater than that of the two-component mixture. The gas handling compressor has sufficient tolerance to enable it to adequately pump this slightly higher density gas mixture.

8.4 DEW POINT MEASUREMENT PROCEDURE

The insulating gas needs to be kept as dry as possible to maintain its high electrical insulating properties. It is common to measure the water dew point as an indirect measurement of the water vapour content in the insulating gas.

The dew point is the temperature to which a gas needs to be cooled for the water vapour to condense into dew or frost. At a given temperature, there is a maximum amount of water vapour that a gas can hold. This maximum amount is called the water vapour saturation pressure. When lowering the temperature of a gas below its water vapour saturation pressure, dew or frost will occur. Typically, a chilled mirror system is used to determine the dew point of the insulating gas. The system consists of following:

— A gas admission system with a manual leak valve and a gas flow regulator (ball type);
— A chamber with an observing window;
— A cooled mirror placed inside the chamber, typically the outside wall of a cylinder that can be filled with a coolant;
— A thermometer to measure the coolant temperature.

Th measurement procedure consists of letting a small flow of gas into the chamber. After several minutes, acetone is inserted inside the mirror cylinder, and dry ice or liquid nitrogen is slowly added into the cylinder. It is helpful to illuminate the mirror with a torch to be able to better observe the formation of the dew. The dew point is the temperature when the light is scattered by the water film condensed on the mirror. Both formation and evaporation points of the water film can be recorded. Once the measurement is finished, the admission valve is closed.

Commercial dew point meters are available which makes the dew point measurement procedure simpler and more reliable.

8.5 SF_6 GAS RECOVERY SYSTEMS

A cost-effective process is to save and recycle the SF_6 insulating gas during tank maintenance. When SF_6 is compressed, it is converted into a liquid and stored in this form. The storage tanks can thus be relatively small. The basic infrastructure needed for a recovery, or gas handling system, is a high throughput vacuum pump, a gas compressor to compress and liquefy the gas, heat exchangers, storage tanks, heaters, recirculating pumps, drying columns, monitoring instrumentation and pipework. Commercially available, portable gas handling plants are available for low volume systems. These portable plants are self-contained and include everything needed for transferring and storing the gas. Large accelerators have a larger volume of gas to handle, and so require a higher level of capability and infrastructure. It is important that a SF_6 sensitive gas detector be used to check for gas leaks during transfers.

8.5.1 Transfer from the accelerator vessel to a storage tank

To remove gas from the accelerator, it is first fed from the accelerator vessel into a compressor. The higher gas pressure within the vessel will provide the driving force to generate the gas flow. The inlet pressure to the compressor is regulated to maintain a constant pressure to ensure effective pumping operation and to minimize the compression time. As gas compression occurs within the compressor, it becomes hot. The hot SF_6 is then passed through a heat exchanger, usually located above the height of the storage vessel, where it cools and converts into a liquid. The liquid SF_6 falls by gravity into a storage vessel. As the pressure within the accelerator vessel reduces to a point where the compressor inlet pressure is too low to adequately maintain compression, a high throughput vacuum pump is switched on. This vacuum pump extracts most of the remaining gas from the accelerator vessel and is operated until the ultimate vacuum pressure is obtained. The small amount of unrecoverable SF_6 inside the accelerator vessel can be sufficient to pose a hazard.

After extraction is complete, the accelerator vessel, vacuum pump, compressor, and storage vessels are isolated, and then air is admitted, backfilling the accelerator vessel. When the pressure inside is at atmospheric, it is important that this initial charge of air be pumped out and vented to atmosphere. Air is backfilled again to atmospheric pressure. This procedure dilutes and flushes any residual SF_6 that might not have been removed by the extraction process. It is important that a suitable SF_6 sensitive detector be used to check for the presence of SF_6 in the bottom of the vessel before personnel entry is made. SF_6, being denser than air, will reside at the bottom of the tank, if present.

8.5.2 Transfer from the storage tank to the accelerator vessel

When the accelerator vessel is ready to be filled with SF_6, all flanges and valves need to first be sealed, and then the air inside the vessel is pumped out. A rotary vacuum pump extracts the air from the accelerator tank and is operated until the ultimate vacuum pressure is reached. Depending on the volume of the tank to be evacuated, the required pumping time can take many hours, if not overnight. It is important that a liquid nitrogen cold trap be used on the inlet of the rotary pump to prevent backstreaming of oil vapours and other volatile contaminants that could occur.

It is essential that as much of the air is removed as possible to reduce any contamination of the SF_6. If there is contamination due to air, the air will not liquefy in the gas compression system during the transfer of SF_6 from the accelerator to the storage tank. The storage tank might not have sufficient capacity to store both the liquid SF_6 and air, possibly causing the maximum pressure rating of the storage tank to be exceeded. Attempts to 'blow-off' the air might result in a significant and expensive loss of SF_6.

Sulphur hexafluoride is transferred into the accelerator vessel as a gas. To begin, the SF_6 is fed in liquid form from the storage tank into a heater where it is converted into a gas. This phase change causes cooling to occur, like in a refrigeration system. The flow of SF_6 into the heater may need to be regulated to sustain an effective change back into a gas. During this operation, the accelerator tank pressure needs to be monitored so that when it reaches its operating value, the heater can be switched off and the accelerator vessel and other gas handling plant valves closed. When most of the liquid SF_6 is removed, the remnants in the storage tank may self-convert to the gaseous phase. It is good practice to have enough unused liquid remaining in the storage tank to maintain a liquid seal over the bottom outlet. This liquid seal will prevent any ingress of air from contaminating the storage tank.

The processes of transferring gases need to be monitored and controlled; an unregulated transfer has the potential to create accidents. To assist with the safe transfer, a log needs to be kept and system conditions periodically recorded. A schematic diagram of the gas handling system and clear written instructions are valuable aids for safely operating the plant.

8.5.3 Transfer of other gases

Nitrogen and CO_2 gases are not liquefied; consequently, they do not have to go through different phases before being transferred to either the storage, or the accelerator vessel. If gas cylinders are used to supply these gases, then the delivery pressure into the accelerator will be regulated at the gas cylinder outlet. The pressure inside the accelerator vessel is monitored and when the desired operating value is reached, the filling procedure is stopped. If the accelerator gas is to be a two-component mix, supplied from separate gas cylinders, then it is prudent beforehand to calculate the number of gas cylinders of each type required to achieve the mixing ratio. Different gases can be stored at different pressures in vendor supplied gas cylinders. Typically, CO_2 is stored at a lower pressure than N_2. The lower pressure gases need to be emptied into the vessel first, followed by the higher pressure gases.

If the mixed gas is being transferred from a storage tank to the accelerator vessel, then the pressure needs to be regulated to ensure that there is minimal pressure shock on associated pipe components. As the full storage tank has a much higher gas pressure than the required operating pressure in the accelerator, the transfer requires no compressor. When transferring from the accelerator vessel back to the high pressure storage tank, then a compressor will be required, and a vacuum pump may be needed later in the process to maintain the necessary inlet pressure on the compressor.

If the three-component gas mixture is to be transferred and stored at high pressure (~15 MPa), the SF_6 will become liquefied by the usual three-stage gas compressor and will deposited as a liquid in the storage tank. When the gas is being transferred back into the accelerator, the SF_6 will remain as a liquid until the storage tank pressure is reduced to below ~2 MPa. This effect can make it difficult to transfer the SF_6 back into the accelerator. Most of the stored nitrogen and carbon dioxide will be transferred into the accelerator tank before the SF_6 liquid/gas transition pressure is reached.

9. WATER COOLING SYSTEMS

Many components of an accelerator system require forced cooling to prevent their operating temperatures rising to unacceptable levels. Internal components of accelerators such as motors and electrical instrumentation dissipate most of their heat load into the insulating gas by conduction and convection. Beamline components such as Faraday cups, beam defining slits and apertures, are heated on contact with the ion beam and usually require water cooling to dissipate their heat load. Other beam transport elements such as deflecting, and quadrupole magnets generate heat from the electrical current passing through their magnet coil windings. Quadrupole magnets are typically forced air cooled while deflecting magnets require water cooling.

The primary type of cooling used in accelerator facilities is chilled water. There are two types of water cooling system: open drain and closed loop. Both systems have their advantages and disadvantages, and applicability. In determining which system would be the most appropriate to install in a facility, numerous issues need to be considered including the required water pressure, flow rate, temperature, operating pressure limits of equipment, electrical resistance of equipment, coolant conductivity limits, complexity, and cost of maintenance.

9.1 OPEN DRAIN COOLING SYSTEMS

Open drain cooling system typically pump water through the cooling lines to the water inlets of the equipment, with the outlet water going to waste, or into a reservoir where it is drawn off and recycled. The flow rate and operating pressure in these systems can be easily optimized and controlled, and backpressure is minimal. In reservoir systems, the water is cooled adjacent to the reservoir, either after the outlet or before the inlet. The most used cooling method is the evaporative water technique using a cooling tower. The advantage of cooling towers is their robustness and relatively simple technology. A disadvantage is the water is exposed to ambient conditions which if not effectively managed, can produce bacterial growth and create health issues such as Legionella, and can require a high level of maintenance to keep the system free from algae and dust. Open drain systems that allow water to run to waste are not appropriate.

Another water cooling method is the refrigeration plant. The refrigeration section replaces the cooling tower and is more efficient at maintaining and controlling the water temperature. These systems require the plant to be monitored regularly for abnormal temperature deviations and will not tolerate becoming dried out. Their main advantage is that they are more compact than cooling towers and can maintain a relatively stable water temperature over a wide range of ambient temperatures. Their main disadvantage is the level of knowledge and expertise required to maintain the plant, which is considerably greater than that for cooling towers.

Both types of open drain water cooling system require the water volume to be regularly monitored and maintained to a sufficient capacity. The make-up water can be adjusted either manually, or automatically using a water float valve or electromechanical valves.

9.2 CLOSED LOOP COOLING SYSTEMS

Closed loop systems are systems where the cooling water is pumped continuously around an enclosed circuit. The temperature of the hold-up volume of water in the closed reservoir is maintained to a preset cool value. Closed loop systems can be prone to overpressure if not properly adjusted. Their main advantages as compared to open loop systems are that the water temperature has a better stability, the water quality can be maintained for long periods without maintenance, and they typically require less make-up water. Their main disadvantage is that system optimization is complex and requires a higher level of technical expertise. The operating pressures of closed loop systems is higher than open drain systems, as the water is being pumped through a loop and back into a closed volume reservoir. Flow rates can be changed either by changing the backpressure while maintaining a constant pump speed, or by changing the pump speed while maintaining a constant backpressure. As a variety of water pressures and flow rates can be produced in closed loop systems, personnel need to take precautions not to create water pressures that exceed the limitations of the equipment being cooled, and not to create flow rates that could cause cavitation.

9.3 WATER SUPPLY AND QUALITY

Preventative maintenance is necessary to ensure long periods of reliable operation. Flexible water cooling lines need to be periodically checked for fatigue and discolouration and replaced as necessary. In the system design, the inside diameter of pipework needs to be large enough to ensure the water flow is sufficient to meet manufacturers specifications, and without increasing a backpressure in the cooling system. Flow meters need to be installed at major equipment outlets to monitor water flow rates, and corrective measures applied if low flow rates are indicated. Flow switches need to be installed on cooled equipment that are temperature sensitive. Remote instrument readouts for the cooling system in the accelerator control room are also important.

Manufacturers of some accelerator equipment specify a low conductivity fluid <1 μS/cm for ion source cooling [21]. This is to provide a high electrical resistance through the fluid and, if water is used, to also control the water chemistry to minimize potential corrosion. It might not be possible to supply low conductivity water to all necessary components in an accelerator facility; instead, one system supplies low conductivity water to the most critical equipment with the remainder being supplied with a lower quality water.

Low conductivity water can be provided using either demineralized or deionized water. Demineralized water can be acidic and may require chemical treatment to keep the pH level near neutral. Chemical treatment additives may also be necessary to reduce algae and for inhibiting corrosion, and periodic flushing of systems may be required to remove scale deposits and dirt buildup inside of cooling lines. Demineralized and deionized water plants can be expensive to buy and maintain and require regular preventative maintenance such as replacing filter and column exchanges. The water in these systems requires periodic testing to ensure adequate water quality is being maintained. When the water quality is not effectively controlled with chemicals, it may be necessary to flush the system with new demineralized or deionized water many times before replenishing the whole system. All make-up water for cooling plants needs to be of least the same quality as the water in service. Every cooling water system requires an effective filtration system that is regularly maintained. Suitable filters are available to remove particulates, algae, and maintain low water conductivities.

9.4 NON-AQUEOUS COOLING

Some components require cooling for which water might not be suitable, such as ion sources which require a high level of electrical insulation. Synthetic fluids such as Syltherm XLT is an alternative cooling medium using a small, self-contained cooling system. This silicon based synthetic fluid has an operating temperature from -100°C to 260°C, and has an extremely low electrical conductivity, far superior to that of water. Table 8 compares the main thermophysical end electrical properties of Syltherm XLT with ultra-pure water at 25°C [22–24].

TABLE 8. PROPERTIES OF SYLTHERM XLT AND ULTRA-PURE WATER

Property	Syltherm XLT	ultra-pure water	units
Electrical conductivity	2.0×10^{-9}	0.055	μS/cm
Thermal conductivity	0.11	0.61	W/m.K
Specific heat	1.8	4.2	kJ/kg.K
Viscosity	1.3	0.89	mPa.s
Density	0.85	1.0	g/cm^3

Syltherm cooling circuits can use a plate heat exchanger with a primary water circuit, typically the same circuit used for magnet cooling. In case of cooling the ion sources, the power requirements are in the range of several hundred watts. Flow regulation of the Syltherm circuit is performed by mechanical valves. The flow rate is set for the maximum cooling requirement of the ion source. If no Syltherm flow is detected, a flow sensor disconnects the ion source power.

10. VACUUM PUMPS, MEASUREMENT AND LEAK DETECTION

Vacuum pumps are used to evacuate gases inside the accelerator tank and beam transport system to create a long mean free path for the ion beam. The mean free path is average distance an ion travels in a gas before it has a collision. In each collision between an ion and a residual gas molecule, the ion is deflected and loses some energy. To minimize the ion beam's angular divergence and energy spread due to these collisions, the mean free path needs to be maximized. In accelerator systems, the ions need to be travel many metres from their point of generation to their destination, so the required mean free path needs to be of least a similar distance. Table 9 lists the calculated mean free path for air at different gas pressures. For a typical low energy accelerator system, a pressure of less than 1 mPa is desirable.

TABLE 9. MEAN FREE PATH FOR AIR VS PRESSURE AT 20ºC

Pressure	Mean free path
1 atmosphere (101 kPa)	1×10^{-7} m
1 kPa	1×10^{-5} m
1 Pa	0.01 m
1 mPa	10 m
0.1 mPa	100 m

When selecting the most appropriate high vacuum pumping system for an accelerator, the following factors need to be considered:

- The ultimate pressure required for normal operations, considering ion beam energies, currents, types of sample to be measured, and the detection systems used;
- Gas load created by inherent residual leaks in the vacuum system, out-gassing of internal components, type of gas to be predominately pumped from the ion sources, and the gas load they create;
- Residual gas composition, and whether backstreaming oil vapours from vacuum pumps may pose problems, such as in radiocarbon dating systems.

The creation of a vacuum usually starts from atmospheric pressure, evacuating the system volume to attain a high or ultra-high vacuum. No single pump can achieve this. A multistage pumping system is used in which a roughing pump (usually rotary vane or scroll pump) is used first to reduce the system pressure to a level when a second pump, a high vacuum pump, can start. The roughing pump is capable of discharging gas to atmosphere and creates the required forepressure for the high vacuum pump to operate. The high vacuum pump is not capable of discharging gas to atmosphere.

The most used types of high vacuum pump on accelerator facilities are turbomolecular vacuum pumps, cryogenic vacuum pumps, and to a much lesser extent, diffusion vacuum pumps. All these vacuum pumps can achieve ultimate pressures of typically 0.01 mPa in optimum conditions.

10.1 ROTARY VANE PUMPS

The rotary vane pump is a widely used roughing pump and serves as a backing pump on high vacuum pumps such as diffusion and turbomolecular. There are single and two-stage rotary pumps. Two-stage pumps are cascaded single stage pumps where the exhaust from the first stage is internally connected to the inlet of the second stage. The extra stage improves the ultimate pressure and throughput of the pump by reducing the back flow of gas and increasing the pumped volume.

An electric motor turns a rotor to which are attached two or three sliding vanes. The rotor is positioned off centre in a cylindrical stator, with the sliding vanes always in contact with inner wall of the stator. The rotor's off centre positioning thus creates different volumes between the vanes. As the rotor turns, the vanes pass through the higher volume side of the offset chamber, the void between the vanes captures a load of gas through an inlet. As it continues to rotate, the vanes move the gas load towards the lower volume side where the gas is compressed. As the vane rotates further, it passes a vent which allows the gas to be exhausted to atmosphere. An oil film provides a seal between the vanes and the stator housing effectively sealing the high vacuum side of the pump from the exhaust side. Oil also surrounds the pump's stator to transfer the heat created by the rotating vanes and gas compression to the pump's exterior casing.

When a gas load is to be pumped containing water vapour, air is admitted into the pumping chamber to help prevent the water vapour from condensing during the compression cycle. Air is admitted through a 'gas ballast' valve and regulated to control the quantity of air admitted. If continual loads of water vapour are pumped, it is important that a cold trap be used in the system to help remove the water vapour before it reaches the rotary pump inlet.

Most rotary pumps have oil level windows on the pump and on the motor end if it is oil cooled. It is important that the oil level and quality is monitored regularly and recharged or changed, as necessary.

As rotary pumps use oil as a lubricant and as a seal, some oil will be pumped through to the exhaust, especially when a high gas load is being pumped. The oil mist leaving the pump can be a hazard if it concentrates in a closed environment. It is important that an oil mist filter is placed on all rotary vane pump exhausts.

10.2 SCROLL PUMPS

Scroll pumps do not use oil as a lubricant or seal and hence their oil-free operation makes them ideal for use with high vacuum pumps where no backstreaming oil vapour can be tolerated. The principle of operation is simple and has been around since the early 1900s. The pump consists of two disks both with spiral shaped cavities as shown in Fig. 23, which are open on one side. One scroll is fixed in position and the other, with its open face in contact with the open face of the fixed scroll, rotates so that the opposing voids become smaller. At the start of the cycle, the scroll collects a volume of gas at the outermost edge. As the scroll rotates, the volume of gas is forced inwards towards the centre

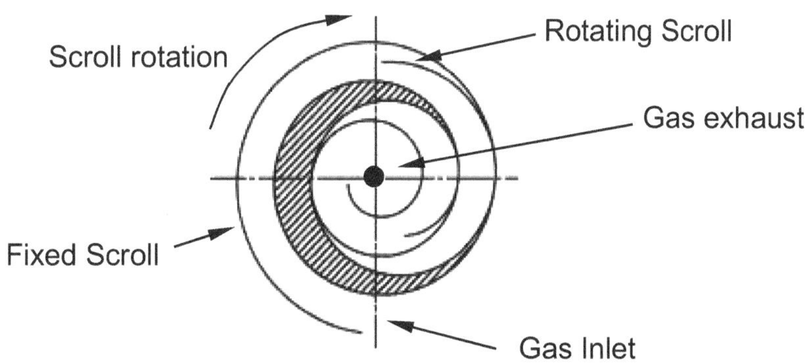

FIG. 23. Operating principle of the scroll pump.

of the spiral where the gas is compressed. The compression continues as the scroll rotates until the gas load reaches the centre where it is exhausted to the atmosphere.

10.3 MEMBRANE PUMPS

Membrane (or diaphragm) pumps are oil-free pumps that work by evacuating the gas from a displacement chamber sealed by a membrane, as shown in Fig. 24.

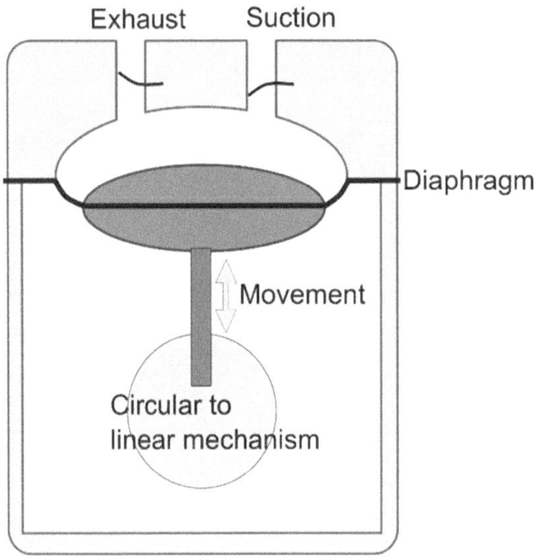

FIG. 24. Schematics of a membrane pump.

The pumping chamber is well sealed by the membrane, so oil is not required for its functioning. The diaphragm needs to be replaced after a long operating time (typically several years) due to cracks that may appear.

A single stage membrane pump creates a vacuum pressure of several 10s of Pa. Typically, the motor is driving two or more stages that are coupled in series to achieve much lower pressures. These multiple stages membrane pumps can be used as primary pumping systems for turbomolecular pumps that have a Holweck stage and can accept ~10 Pa to ~100 Pa at their exhaust.

A drawback of membrane pumps is their relative low pumping speed (several m^3/hr). In case of large systems, a preliminary roughing pump or another type of fore-vacuum pump may be required. An advantage of membrane pumps is their power consumption, being approximately one order of magnitude less than the oil or scroll pumps.

10.4 TURBOMOLECULAR VACUUM PUMPS

Turbomolecular pumps operate using the turbine principle of compressing gases. Rotor blades connected to a rotating shaft move over fixed blades attached to the wall of the pump body. The spacing between the blades is very small. The rotor revolves at thousands of rpm, with some designs exceeding 30 000 rpm. A turbomolecular pump will have many layers of rotors and stators to create a high pumping speed. A brief description of the internal parts of a typical molecular pump can be found in Ref. [25].

A gas molecule entering the pump inlet collides with a rotating blade, and due to the pitch angle set on each blade, the gas molecule is knocked downwards into the pump. It will next hit a fixed blade, which is also angled to knock the gas molecule further downwards. When it hits the next rotating blade, it is usually knocked further down again. In the event of the molecule bouncing upwards, it will most likely hit a fixed or rotor blade above and be knocked back down. The further the molecule moves down through the layers of rotors and stators, the less chance it has of returning out of the pump. This process continues until the molecule reaches the outlet of the pump, where it is removed by a rotary backing pump.

Turbomolecular pumps are the most used high vacuum pumps on electrostatic accelerators. Their ability to withstand surges in gas load, pump away gases contaminated by oil vapours, and their relatively high pumping speeds for light gases such as hydrogen and helium make them the preferred high vacuum pump. Turbomolecular pumps have an inlet to outlet compression ratio from 10^3 to 10^6, depending on the design. Their rotary backing pumps have typically, an ultimate inlet pressure from 1 Pa to 0.1 Pa. Using the above compression ratios, the ultimate pressure at the inlet of a turbomolecular pump is between 1 mPa and 0.001 mPa.

Over an extended period of operation, oil vapours from the rotary backing pump can migrate through the turbine section of the turbomolecular pump and into the high vacuum system being pumped. This backstreaming of vapours is caused by a combination of wall creep and vapour transport. To mitigate this effect, a liquid nitrogen cold trap or other foreline trap can be placed between the turbomolecular and rotary pump, or a liquid nitrogen cold trap at the inlet of the turbomolecular pump. The latter arrangement has the additional advantage of the liquid nitrogen cold trap freezing out condensables, such as water vapour, thereby increasing the effective system pumping speed.

Turbomolecular pumps can have either water or air cooling. Most modern pumps have air cooling, which is provided by a fan mounted over the base of the pump. Air is blown over the lower bearing area of the pump and along the pump body to cool it. When a water cooled pump is used, it is important to ensure that there is an adequate water flow before the pump is started. It is important to interlock the pump with a water flow switch so that accidental startup without water flow is prevented. Most modern pumping systems have controllers that interlock and protect the pump during pumping operations. On water cooled pumps, it is important to supply low conductivity water, less than 100 μS/cm, to minimize corrosion effects over the lifetime of the pump.

It is important that turbomolecular pumps be left running continuously, if possible. Experience has shown that uninterrupted operation of the pumps extends their lifetime, compared to pumps that are cycled on and off between uses. During pump shutdown, it is important to allow the rotor speed to decrease naturally to a stop before venting the pump to air. A forced stop can raise the rotor to a high temperature, sufficient to evaporate condensed oils near the outlet section of the pump and promote their transport upwards through the pump into the much cleaner inlet section.

Some of the early model turbomolecular pumps, and most single-ended turbomolecular pumps, can be damaged by large gas surges. The fast-spinning rotors can bend due to the high gas pressures pushing on the blades and can cause the rotors to contact the fixed stators causing a catastrophic pump failure in which blades break away from their mounts, impact onto the fast spinning rotor ramming broken blades onto the walls of the pump body and other internal components. The pump is destroyed and is irreparable.

During the past few decades, many pump manufactures added an additional stage to the turbomolecular pumps to allow higher exhaust pressures. The most common one is the Holweck stage, which consists in a helical type of channel (like the Archimedes screw in its operating principle). Using this stage allows operating with exhaust pressures up to 2 kPa.

10.5 CRYOGENIC VACUUM PUMPS

Cryogenic vacuum pumps operate like refrigeration compressors that compress ultra-high purity helium gas to pressures of about 1500 kPa and then allow it to expand, cooling two separate stages. Figure 25 illustrates the main components of a typical cryogenic pump system.

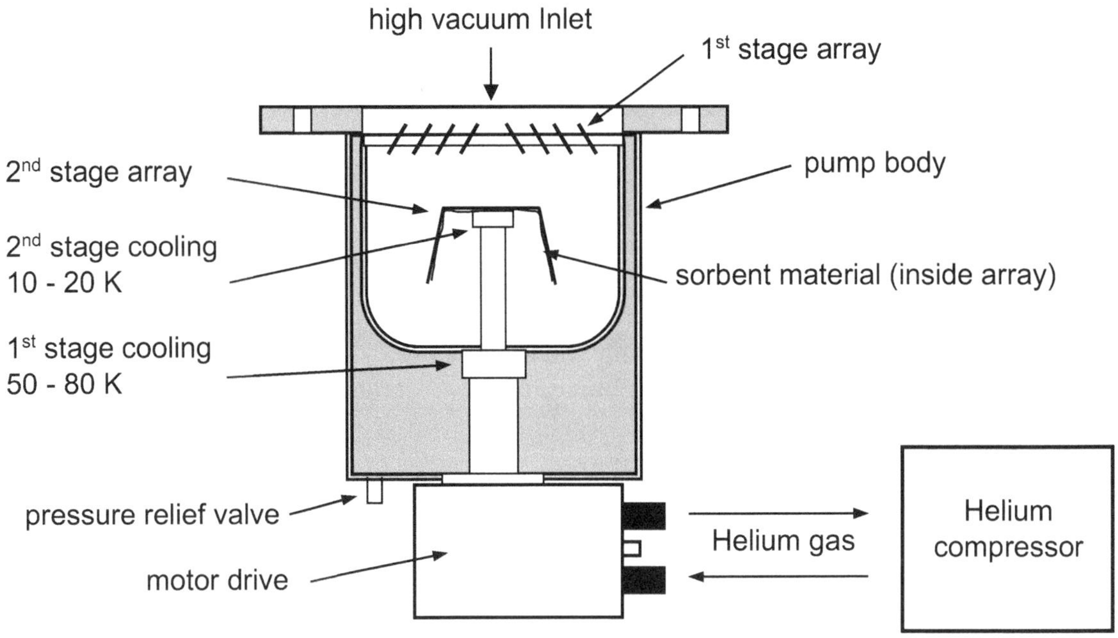

FIG. 25. Main components of a typical cryopump system.

The pump is made of a metal cylinder that contains cooled pumping arrays and underneath, a cold head and helium compressor. When the compressed helium gas expands, it becomes cold producing a refrigeration action. The first stage cools to a temperature of about 77 K and serves to capture water vapour and other high temperature condensables from the vacuum system. The second stage of the cold head reaches a temperature of about 20 K, which sufficiently low to capture most residual gases found in a typical vacuum system. Hydrogen and helium cannot be captured as they require colder temperatures to become liquefied. A layer of activated charcoal is bonded to the second stage pumping array to absorb these gases. Helium gas is transferred between the cold head and the pumping unit is via stainless steel convoluted tubing. Each end of the tubes has a shut off valve (self-sealing tubes) to ensure that air does not get into the system during tube connections or disconnections.

The main limitation of cryogenic pumps is their inability to have high pumping speeds for helium and hydrogen gas, and the inability to maintain high pumping speeds where excessive backstreaming oil vapours or moisture is present. These pumps work on the principle of freezing out gas molecules at temperatures just above liquid helium temperature, hence their inability to effectively pump helium or hydrogen gas. The pumps are effective at pumping atmospheric composition gases, but they eventually become saturated in their storage capacity and so the pump loses its capacity to pump more gas. When using cryogenic pumps, the vacuum system and pump first need to be evacuated with a backing pump, preferably with a cold trap, to a pressure of at least 0.1 Pa. Once achieved, the backing pump can be isolated, and the cryogenic pump can be activated and opened to the vacuum system.

10.5.1 Cryopump maintenance

Cryopumps require little routine maintenance. The only regular service required is the replacement of the high purity filter in the compressor called the absorber. This service requires the compressor to be shut off and the cover removed. The absorber attachments (screws) to the compressor frame are removed and the self-sealing couplings of the inlet and outlet helium lines are taken apart. A new or refurbished absorber is installed using the reverse procedure. If the static helium charge in the system falls below the minimum recommended by the manufacturer, sufficient helium needs to be added to the compressor to restore it to the normal operating pressure. If helium losses are high (e.g. requiring recharging more than once per year), it is likely that there is a helium leak somewhere in the system. Leak testing of the self-sealing joints in the compressor, the hoses, and other welded or threaded joints needs to be undertaken.

10.5.1.1 Ratcheting

'Ratcheting' is a noise sounding like that of a ratchet being used, that the cold head expander makes when air, moisture or other materials contaminate the closed helium loop of the cryopumping system. In the early stages of helium contamination, the sound made is usually like a click or knock, and if contamination is left to build, the sound develops into a loud hammering noise. The sound is generated by the contaminants freezing out in the expander of the cryopump cold head and interfering with the movement of the expander piston. The pump may become damaged if left for long periods without corrective maintenance.

The pump may be returned to normal operation in-situ. First, the pumping unit needs to be isolated, turned off, and allowed to warm to ambient temperature. Helium is purged through the helium pumping system with new ultra-high purity helium by venting and refilling it several times. This will dilute the contaminants. Then the system is recharged to normal static pressure with new ultra-high purity helium. Before restarting the cryopump, it is important that the pump is also regenerated. Manufacturer service manuals may recommend other methods of rectifying ratcheting.

10.5.1.2 Poor ultimate vacuum

Cryopumps pump gases by capturing and trapping the gas molecules. The pumps have a finite capacity for pumping gases and eventually become saturated leading to a reduction in their ability to pump gas out of the vacuum system. The first indication of this occurring is when the temperature or pressure gauge on the cold head begins to rise, followed by a rise in the vacuum system base pressure. If the base pressure has risen to an unacceptable level, then the cryopump needs to be removed from service and regenerated. Regeneration removes the gas load that is frozen on the arrays or trapped in the activated charcoal, producing a clean and empty system capable of pumping a new gas load.

Regeneration is a straightforward procedure and only requires a roughing pump with a cold trap. If a roughing pump fitted with a cold trap is not available, then another high vacuum pumping system may be used instead to pump out the stored gas load. The regeneration procedure can take about 12 hours, from when the cryopump is taken offline until it is back online. The procedure is as follows:

- The cryopumping unit needs to be first isolated from the vacuum system, turned off, and allowed to warm to ambient temperature. As the warming may take more than 8 hours, it is usually best to allow warming overnight.
- When warm, the cryopump needs to be pumped through a cold trap by a backing pump until a base pressure of at least 0.1 Pa or better is achieved.

— The backing pump needs to pump the cryopump for at least 2 hours to remove as much of the stored gas load as possible.
— When pumping is completed, isolate the cryopump from the roughing pump and turn on the cold head. Leave for about 3 hours or until the cold head has cooled to near its normal operating temperature.
— When cold, the pump can be returned to service.

10.5.1.3 Housing sweating

Cryopumps may begin to 'sweat' i.e. develops water droplets on the outside of the pump housing or even to form a layer of frost. This is due to the outside of the cryopump become cold and condensing water vapour in the ambient atmosphere. It is a sign of poor thermal insulation between the outside of the cryopump and its very cold interior surfaces, and is caused by either a poor vacuum inside, or a build-up of ice inside of the pump forming a heat conducting bridge to its exterior. The pump's storage capacity has become saturated and is unable to effectively pump further. Rectification is by regenerating the cryopump,

10.6 DIFFUSION PUMPS

Diffusion pumps have been used on accelerators since the early years of accelerator operations. They are effective pumps, have no moving parts, and servicing is relatively easy. They require a backing pump on the outlet stage to remove the exhaust gases and to maintain a low backpressure. Figure 26 shows the internal arrangement of a diffusion pump.

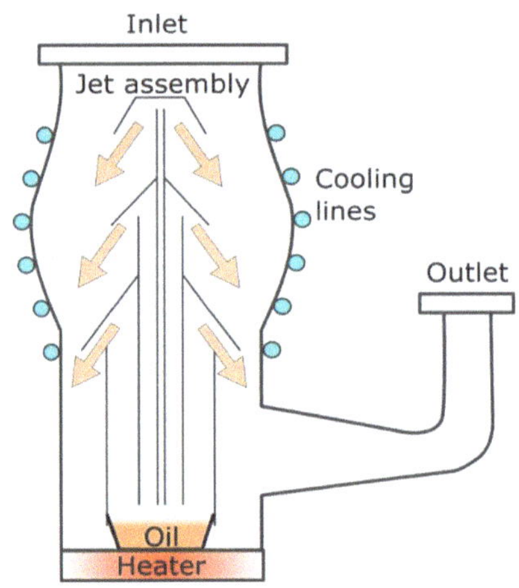

FIG. 26. Internal arrangement of a diffusion pump.

A diffusion pump is effectively a cylinder with an electric heater underneath and an upside-down multi-funnel shaped chimney on the inside. The pumping medium used is oil, possessing a very low vapour pressure. When the heater heats the bottom of the pump, the oil becomes hot and begins to vaporize, rising through the inside of the chimney. The molecules in the vapour emerge through ring-shaped nozzles at the top of the chimney at supersonic speed. They are directed downwards toward the cooled pump wall where condensation takes place. On contact with the wall, the vapour condenses and flows back to the bottom of the pump where it is heated again to form vapour again and the cycle continues.

Gas molecules entering the pump are mixed with the larger oil molecules moving downwards at high velocity towards the pump wall and compress in the bottom stage of the pump where they are removed by a roughing pump.

For effective operation, the mean free path of the oil molecules needs to be greater than the distance from the nozzle to the wall of the pump. In addition, the pumped gases that accumulate at the bottom stage exhaust passage need to be removed to prevent disintegration of the lowest vapour jet stream. If disintegrated, the pumping speed declines due to the velocity of the air molecules in the upward direction becoming the same as the downward direction. Therefore, the attached roughing pump needs to be capable of providing a high throughput with a backpressure of better than 10 Pa. It is important that a cold trap is placed between the diffusion and the roughing pump to stop backstreaming oil mixing with the diffusion pump oil, otherwise the smaller sized molecules of rotary pump oil will eventually saturate the larger molecule diffusion pump oil severely reducing its pumping speed.

Diffusion pumps, such as turbomolecular pumps, require cooled water to keep the inner casing cool. It is important that a flow switch is connected to the cooling line to shut the pump down in the event of loss of water flow, and a thermal cut-out switch on the heater element. Low conductivity (< 100 μS/cm) deionized water is needed to help inhibit corrosion.

Various types of fluid are used in diffusion pumps. The most common types are mineral oils, silicon oils, and pentaphenyl ether. Table 10 lists some comparative properties of these fluids. Mineral oils are suitable for general applications down to a pressure of ~10^{-4} Pa. Silicon oils are very stable compounds at high temperatures and will provide an ultimate pressure between 10^{-2} Pa and 10^{-6} Pa. Pentaphenyl ether has an extremely low vapour pressure and these oils are the most thermally and chemically stable diffusion pump oil.

TABLE 10. TECHNICAL DATA FOR DIFFUSION PUMP FLUIDS

Property	Mineral oils (Balzers)		Silicon oils		Pentaphenylaether
	61	71	DC704	AN175	SANTOVAC5
Theoretical vapour pressure (Pa)	2×10^{-5}	2×10^{-6}	2×10^{-6}	4×10^{-8}	1×10^{-8}
Viscosity at 25°C (mm^2/s)	171	410	39	175	1000
Chemical resistance	good	good	better	better	best
Thermal resistance	good	good	better	better	very good
Pressure range (Pa)	1 to 10^{-4}	0.1 to 10^{-5}	0.1 to 10^{-5}	10^{-3} to 10^{-6}	0.1 to 10^{-6}

During regular use, backstreaming oil from the diffusion pump can slowly migrate into the vacuum system and deposit on the inner surface of the accelerator tubes degrading their high voltage hold-off properties. This backstreaming can be minimized by using diffusion pump oils with an extremely low vapour pressure, and using water cooled baffles and/or liquid nitrogen cold traps on the diffusion pump inlet port. Hot diffusion pumps need not to be exposed to air at ambient pressure as the pump oil will quicky react and deteriorate and the pump will lose its vacuum pumping ability. Diffusion pumps need to be allowed to cool down to ambient temperatures before venting to air.

10.7 ION PUMPS

Ion pumps work by ionizing residual gas molecules in the vacuum, accelerating them to high speeds to impact on a cathode material which sputters, capturing and trapping the ionized gas by various processes. There are numerous designs for ion pumps; the simplest consisting of a stainless steel box, surrounded by a very strong permanent magnet, and inside the box is an electrically isolated anode assembly of numerous open-ended cylindrical tubes. A grounded rectangular titanium cathode plate is located near each open end of the anode assembly.

To start up, the vacuum pressure needs to be reduced to ~0.01 Pa or better, then a voltage of ~7 kV is applied between the anode assembly and the titanium cathodes. Free electrons are attracted to the anode but are trapped by the magnetic field, causing them to rotate until they eventually collide with a gas molecule ionizing it. The positively ionized gas molecule is accelerated towards the titanium cathode where it makes an impact sputtering some titanium atoms. If the gas molecule is a reactive gas, such as hydrogen, oxygen or nitrogen, a chemical reaction occurs forming sputtered molecules of titanium hydrides, titanium oxides or titanium nitrides, which get deposited on the walls of the anode assembly.

This chemical reaction process cannot occur for noble gases. Instead, the titanium cathode is replaced by a tantalum cathode. The impacting inert gas ion is converted to a neutral atom and is reflected from the tantalum and gets adsorbed on the anode surface. Eventually it becomes buried beneath a layer of sputtered tantalum.

Ion pumps can have capacities up to about 7000 l/s with an ultimate pressure of $\sim 10^{-8}$ Pa. The full pumping speed of these pumps is typically developed in the pressure range from $\sim 10^{-5}$ Pa to $\sim 10^{-7}$ Pa. Ion pumps sometimes can have instabilities when pumping inert gases. To improve performance and stability for noble gases, other internal designs such as triode structure are used. The lifetime of ion pumps is usually determined by amount of sputtered cathode material and condition of the cathode. Sputtered cathode material coats surface such as insulators which results in the ion pump shorting out, and replacement of components is required. Heavily sputtered cathodes can be eroded through, requiring the cathode to be replaced.

10.8 COLD TRAPS

Cold traps have a cold surface which is exposed to the vacuum to freeze out contaminants such as residual gases, oil, and water vapour, thus improving effective pumping speeds and base vacuums. The removal of these contaminants protects critical surfaces in the electrostatic accelerator and other equipment inside the vacuum system. Cold traps are typically located at strategic positions to maximize their effectiveness: between rotary and high vacuum pumps, on the high vacuum side of high vacuum pumping systems, on beamlines, and near end stations. They block backstreaming oil and other contaminants from vacuum pumps and other sources to provide a clean vacuum for accelerator operations and analytical instruments.

The liquid nitrogen (LN_2) trap is the most common type used. It consists of a stainless steel construction with an inner well (reservoir) that is filled with LN_2. The inner well surfaces cool to near 77 K and is in effect a static cryopump. Some special types have coated inner surfaces such as gold to reduce outgassing. Cold traps over time eventually become loaded with frozen out vapours and their capture efficiency declines. It is important that periodic cleaning is done at least every 6 months. Cold traps come in many geometries including upright cylinders, horizontal cylinders, ball type, cylinders with centre holes for mounting in beamlines, and baffle types where the cooled fins have a high surface area exposed to the vacuum.

Refrigeration type cold traps use a small refrigeration unit to cool a cold finger that is inserted into the vacuum. Unlike the immediate response from well type LN_2 cold traps, the refrigeration units take ~15 minutes to cool down, so the user needs to ensure it is switched on before its use is required.

These units can reach temperatures of better than -100°C which is sufficient to freeze out most oil, water, and other vapours.

To achieve a long useable time from cold traps, they need to not be used in vacuums with a pressure above 1 Pa, otherwise the high gas load will quickly saturate the cold surface. The cold trap needs not to be made cold until the vacuum space inside the cold trap has been roughed out with a roughing pump.

To physically clean heavy deposition on the trap, it needs to be done with appropriate personal protective equipment including gloves and safety glasses, as the trapped contaminants may be harmful. The procedure is as follows:

- First, isolate the trap and allow it to warm to ambient temperature. To speed up the warming when the LN_2 has nearly boiled dry, hot air can be gently blown into the well with a hot air gun, or by inserting a heater connected to a power supply via an isolation transformer.
- When at ambient temperature, remove the well (reservoir) from the system and wash off the accumulated contaminants from the well walls with solvents such as methyl alcohol and acetone, and then blow dry with hot air. If the trap is likely to have only light deposits, the trap can be warmed in-situ with a hot air gun or heater while being pumped by another vacuum pumping system. The heat will warm the well walls, evaporating any trapped gases and water vapour which will be pumped away by the other pumping system.
- Before returning to service, it is important that the inner well of cold traps be blown dry with compressed air or dry nitrogen to expel any residual moisture that may be present.

10.9 ISOLATION VALVES

Isolation valves are an essential part of any vacuum system. They are used to isolate and protect a vacuum system and vacuum components against power failures, abnormal and transient vacuum events, to isolate sections for maintenance, and to isolate detection systems and analytical chambers when they are cycled to atmosphere to change samples.

Many valves used on electrostatic accelerators seal only on one surface, and therefore have a particular orientation for installation. The surface to which the seal makes contact is the high vacuum side of the valve. It needs to face the side of the vacuum system most likely to be under vacuum when the valve is closed or the side with the highest vacuum. If the valve is not oriented correctly, the valve seal may be lifted off the seal face by the differential vacuum pressure.

Isolation valves come in a variety of configurations, with designs to suit a particular function and be cost-effective for that function. Commonly found isolation valves on accelerator vacuum systems include gate valves, butterfly or baffle valves, needle or metering valves, shut-off valves, and air admittance valves.

Gate valves have a large, unobstructed, and high throughput throat through the centre of the valve, which makes them ideal for use in beam lines, analysis chambers and detection systems. The method of actuation ranges from manual lever type action, pneumatic action, electric actuators, and electro-pneumatic. When an electro-pneumatic gate valve is coupled with a vacuum control system, it enables automatic vacuum protection against abnormal and transient vacuum excursions. For critical locations which need to be isolated as fast as possible, fast-acting gate valves are needed. When a vacuum excursion occurs, there is a time lag between when the vacuum controller actuates the gate valve to when it becomes fully closed. To minimize this time lag, fast-acting gate valves can be utilized in which their time to close is much shorter than a standard gate valve.

Butterfly or baffle valves are mostly used with diffusion pumps as they have a low profile and fit neatly above the pump. Care needs to be taken to keep the O-ring seals clean as the baffle typically sits vertically, exposed to the vacuum system and potential contaminants. They normally open through a manual quarter swing lever action, but other types of closure actuating mechanism using electro/mechanical/pneumatic are available.

Needle or metering valves are used for the controlled admittance of gas and are screw operated. Common uses include regulating the amount of gas admitted to ion sources, gaseous detection systems, and to restrict a vacuum pumping system when evacuating equipment containing sensitive or fragile vacuum windows.

Shut-off valves are used where a controlled opening or closure is required to reduce pressure surges. They are mostly manually operated. Electro-pneumatic types are sometimes used between roughing and high vacuum pumps and are wired to close when power is removed, protecting the system from backstreaming oil vapours. Most types have a stainless steel bellows that separate the valve stem from the vacuum, ensuring leak free operation.

Air admittance valves admit air into pumps and components after shutdown. Rotary pumps need to be brought to atmospheric pressure when not in use to stop oil being sucked back into the inlet tubing. They operate manually or electrically. If electrically operated, they will be normally open valves that are wired so that they are only closed when the vacuum pumps are operating. Once the pump is turned off, the valve opens venting the system.

Valve seals range from various rubbers to metal and plastics. Viton is an excellent rubber for valves as it has good elastic recovery and low outgassing rate. Small metal particles can be created over time from valve operations and can deposit on the seal material. They are often unseen until sufficient buildup causes a leak to occur. All seals need to be inspected during vacuum maintenance and replaced if necessary.

It is important that vacuum systems are vented with an inert gas or dry nitrogen. This will make pump down after cycling quicker and will help to maintain a clean system free from water vapour and atmospheric particulates.

10.10 VACUUM MEASUREMENT

Vacuum pressure measurement can be divided into three categories: low, high, and ultra-high. There are many different techniques used to measure vacuum pressure in these categories, however no one gauge can cover them all. By using several different types of gauge, spanning different pressure ranges, it is possible to measure vacuum system pressure from atmospheric pressure to 10^{-8} Pa or lower. Table 11 lists the main types of vacuum measurement gauge and their effective measurement range.

TABLE 11. TYPES AND MEASUREMENT RANGES OF COMMON VACUUM GAUGES

Vacuum pressure gauge type	Pressure Range	Typical Use
Mechanical	100 kPa to 100 Pa	From near atmospheric pressure to roughing pressure. Good as an indicator of vacuum presence
Thermal conductivity	100 Pa to 0.01 Pa	Excellent in the 'roughing range' on roughing pumps and in vacuum systems where low vacuums are regularly used
Ionization	1 Pa to 10^{-8} Pa	High to ultra-high vacuum systems

10.10.1Mechanical type gauges

Mechanical gauges obtain their pressure indication from the physical movement or physical change of components.

A Bourdon gauge comprises a hollow, coiled metal tube. When the pressure inside the tube is different to atmospheric, the pressure difference creates a stress on the tube which deforms, creating a mechanical force to move an indicator needle. As the mechanical movement can be created by either positive or negative pressure, Bourdon gauges can measure pressures above or below atmospheric. They are typically used on pressurized systems such as gas cylinders, or gas storage tanks, as an indicator of gas quantity.

The piezo gauge is an electronic gauge head comprising a small ceramic element that when subject to a mechanical stress, generates an electrical voltage, which is then amplified and displayed. They are relatively inexpensive and are simple to operate and have a measuring range of 100 Pa to 100 kPa, making it suited for positive to negative pressure applications. Some types of piezo gauge head can be difficult to recalibrate when they become old and display erratic pressure readings. Some models are sensitive to nearby high voltages so care needs to be taken where they are used. As they function by mechanical stress, they are independent of the gas composition.

10.10.2 Thermal conductivity gauges

Thermal conductivity gauges are one of the most common gauges used for measuring low vacuum. They have a heated filament that, in the presence of gas, gets cooled due to the heat being conducted away by the gas. The lower the gas pressure, the less the cooling effect. A Pirani gauge operates by having a constant temperature filament, with the filament current changed to maintain the temperature. The change in current (i.e. the change in filament resistance) is the indicator of pressure. A thermocouple gauge has a constant current in the filament, with a thermocouple which measures its temperature. The change in temperature is the indicator of pressure. As different gases have different thermal conductivities, the pressure reading is dependent on the type of gas. The gauge controller needs to be calibrated for the type of gas to measured, with the default set to normal atmospheric composition.

Thermocouple gauges only work reliably in the pressure range from about 1000 Pa to 0.01 Pa. There is no maintenance associated with these gauges other than keeping them clean. A solvent such as ethanol can be used to rinse out trapped contaminants, however this might not be effective as the contaminants may have corroded the filament. It is usually best to replace the gauge if problems persist.

10.10.3 Ionization gauges

Ionization gauges are either a hot cathode gauge, which has a hot filament creating the ionizing electrons, or a cold cathode gauge without a filament, which produces ionizing electrons as part of a self-sustaining discharge. In both cases, when the electrons collide with gas atoms or molecules, they produce positive ions which are collected creating an ion current which proportional to the gas pressure. As with thermocouple gauges, a calibration needs to be used for different gases in the vacuum system as the ionization cross-section depends on the gas species.

In a hot cathode gauge, an electrically heated filament emits electrons that are accelerated towards a more positive grid surrounding the filament. The accelerated electrons collide with gas molecules in the volume between, producing positive ions that are attracted to a collector electrode from which the flowing ion current is measured.

These gauges are simple in design and have a measuring range from about 10^{-2} Pa to 10^{-6} Pa with an accuracy about ±30%. Lower pressure readings are not possible in standard designs as the primary ionizing electrons create soft X rays which can produce photoelectrons, adding an extra unwanted current source in the collector electrode. The filaments are prone to burnout in the presence of higher pressures, so an interlock is needed to turn off the gauge power at higher pressures.

Cold cathode gauges, having no filament, rely on a stray field emission electron or electron produced by an external event such as a cosmic ray to generate starting electrons. This may take some minutes. Once starting electrons are generated, a magnetic field confines the electrons to long helical paths, ionizing the gas in the vacuum creating a self-sustaining discharge. The collected positive ion current is proportional to the gas pressure. The measuring pressure range is about 10^{-1} Pa to 10^{-5} Pa with direct readout, and about 10^{-1} Pa to 10^{-11} Pa with a high precision current amplifier. They are sensitive to contamination caused by the ionizing gases and are gas dependent. Their robust design allows for physical and chemical cleaning.

10.10.4 Vacuum gauge maintenance

Vacuum measuring equipment needs to be of good quality, have a high reliability, and operate across the expected range of the accelerator vacuum systems. Inadequate and substandard vacuum measuring equipment invariably leads to system failures, unnecessary damage to acceleration tubes and other high voltage equipment, and extended maintenance periods. It is important that gauges and controllers are standardized where possible, and that only types that can be easily serviced and replaced are used. This allows for cross-checking using like-for-like gauges and swapping of gauges when a spare gauge head is unavailable.

Most modern vacuum gauges have built-in protection against operating at high pressures, however many old systems do not. Power isolation interlocks need to be considered to isolate and protect high voltage gauges when pressures approaching their recommended maximum safe operating pressure is detected.

10.11 LEAK DETECTION

The pressure in a vacuum system decreases exponentially with time after the pumping system is switched on, with the ultimate pressure determined by the gas load to be pumped and the vacuum pump's pumping speed. The gas load consists of outgassing of materials, process gases such as in ion sources, and leaks. No system is leak free. Leaks are usually indicated when either the ultimate pressure is higher than normally achieved, or the pressure decreases at a slower rate than normal while the system is being evacuated. Vacuum leaks are often small and usually cannot be located without leak detection equipment such as a dedicated helium leak detector or a residual gas analyser (RGA).

A suspected leak might not be a real leak. An extra gas load may be the result of outgassing from a recently introduced change or modification to materials used, or a vacuum pump not pumping as efficiently as usual. To help locate the source of an extra gas load, isolate the vacuum system from all pumps and monitor the rate of pressure increase. If it is relatively fast, the source is likely to be internal to the vacuum system. If it is relatively slow, the source is likely to be external, such as the vacuum pumps. Only operational experience will decide what is fast and what is slow.

10.11.1.1 Basic leak detection methods

A vacuum leak in an old system can arise from ageing seals and stressed components, and locating the leak can sometimes lead to more leaks being created. In new systems, finding a leak can be easier as components are less prone to damage from disassembling and reassembling. When locating a leak, sections of the vacuum system are isolated and use the existing gauging to look for pressure rises in the isolated sections. The leak is usually in the section with the fastest rate of pressure rise. When isolating sections of the system, it is prudent to use existing isolation valves however if this becomes inadequate, then blank flanges are an acceptable alternative.

When finding leaks using helium and RGA leak detectors, small leaks can be missed because of a time delay between the leak test gas entering the leak and the detector indicating a leak. When the gas nozzle is spraying gas at a leak, the gas after entering the leak may have a long travel path to the gas detector head in the leak detector and may cause delays of many seconds (or longer) before any response is indicated by the detector. Thus, it is important to isolate the vacuum system into small sections where possible, and only test small areas at a time, waiting some tens of seconds or so for a detector response before moving on to the next test location.

Leak test gas pressures need to be low so that there is control over the area where the gas can flow. If the pressure is too high, the gas may disperse away from the test area and into a leak far removed from the area being tested, giving a false location by the leak tester. The use of a gas gun with a regulating trigger is the most convenient method to dispense gas for leak testing. If the detector has an audible output, this can be useful to use where the leak detector cannot be brought close to the study area. If an audible output is used and a small rise in pitch is heard, then check that this is not an excursion in the stability of the detection head before judging it as a leak. Sometimes the audible signal can naturally vary in intensity, falsely indicating to the user into thinking that there may be a small leak.

When a small leak has been isolated to a particular area, then this area can be subdivided and made smaller. In this way, the leak area is reduced, and the precise location of the leak is more likely to be found. In addition, the leak test gas pressure can be reduced so that an even smaller area is exposed to gas. The area around the suspected leak can be masked to improve resolution of location. When the leak location has been presumed identified, it is useful to stop the flow of gas and leak detection for a few minutes to allow a quiescent condition to establish, then recheck for confirmation. This will help to avoid a false positive.

10.11.1.2 Leak detectors

Leak detectors are either helium or RGA types and have many similarities. They are both mass spectrometers. A helium leak detector is tuned to detect only helium and has its own mobile vacuum pumping system. Helium is readily available and as it has a small atomic size, it can readily penetrate the smallest leaks. A RGA can detect gases other than helium. This can be an advantage in certain situations where the ability to identify residual gases in the vacuum system, such as hydrogen, water vapour, solvents, and molecules released from the outgassing of materials. A disadvantage of RGA is that it does not come with its own vacuum pumping system. The RGA detection head is inserted into the vacuum system, typically via a ConFlat flange.

10.11.1.3 Using vacuum gauges for leak testing

Vacuum gauges can be used to find large leaks, leaks such that vacuum pressures cannot be lowered sufficiently enough to enable helium or RGA leak detectors to operate. This method senses pressure rises from solvent vapours penetrating through the leak. The procedure is to spray solvents such as ethanol, methanol, isopropanol or acetone, onto the suspected leak and observe any pressure rise on the vacuum gauge indicating a leak due to the evaporating solvent entering the vacuum system. The sprayed solvents can penetrate large leaks. The pressure may also fall, as the evaporating solvent vapours may cause local freezing, temporarily sealing the leak. Acetone needs to be used with caution as it is harmful to Viton O-rings.

10.11.1.4 Leak checking acceleration tubes using a residual gas analyser

An RGA can be used to determine if an accelerator tube has a leak without entering the accelerator pressure vessel. The RGA is tuned to look for the insulating gas, SF_6, CO_2 or N_2 which will be detected if there is a leak in the accelerating tubes. The RGA head is placed in the beamline as close to the accelerator vessel as possible but with an isolation valve between the head and the pressure vessel. A base measurement is first taken, then the isolation valve is closed and if the level of the insulating gas being detected reduces, then there may be a leak through the accelerator tubes. This then gives reason to enter the vessel to search further for the leak.

10.12 VACUUM SEALS

Most ultra-high vacuum seals are made from copper, aluminium, or other soft metal. When properly installed, they have extremely low leak rates and are maintenance free for long periods of operation. However, they can be used only once. A new metal seal need to be installed every time the vacuum connection is disassembled.

Rubber or plastic seals are reusable however, they can leak when old. A maintenance programme is encouraged for regular replacement of rubber and plastic seals. Some rubber seals will eventually harden and become brittle. Many older vacuum systems may have used natural rubbers or rubbers that were developed for mechanical plant and equipment rather than static vacuum systems. Viton is the most common rubber material that vacuum seals are made of and is bakeable up to about 150°C.

When selecting seals of any type, it is necessary to ensure they are the correct size. Stretching a rubber O-ring to make it fit a larger sized flange is not appropriate as it may lead to premature cracking or poor sealing. Using vacuum greases to compensate for poor sealing properties is to be avoided. It is best to use no grease at all as greases can trap dirt and dust creating a leak point.

There are three main types of vacuum sealing system used on accelerators and associated equipment: Dependex, ConFlat, and ISO.

10.12.1 Dependex

The Dependex system uses a 5° sloped sealing surface for a rubber O-ring which is clamped together with slip-on flanges retained by large circlips. The angled Dependex seal reduces the risk of the seal being scratched when placed onto a rough surface. The Dependex style is only suitable for high vacuums. They are inexpensive and moderately easy to manufacture from aluminium tubing.

10.12.2 ConFlat

ConFlat (or CF) is the most widely used ultra-high vacuum type flanging. The CF style uses a recessed angled sealing edge, which penetrates a soft metal gasket when tightened. The recessed edge stops the sealing surface from becoming scratched if it is placed onto a rough surface. The main type of metal gasket used is made from annealed copper gasket. Only high tensile bolts need to be used with these flanges. It is important that the CF flanges are tightened in a sequential pattern so that the gasket seal is penetrated evenly. An alternate or opposite pattern can lead to distortions in the gasket and cause a leak. CF flanges are usually made from stainless steel and are relatively expensive to purchase.

10.12.3 ISO Large Flange

The ISO standard range is the least common type used on accelerator beam line flanges. However, they are common on vacuum components. This style is only suitable for high vacuums. One of the major problems with ISO Large Flange (LF) type fittings is that some types seal on a flat surface, and so are susceptible to scratching if inappropriately placed onto a rough surface. ISO LF seals come in one of two different clamping styles, K and F. ISO-K flanges are joined with double-claw clamps, which clamp to a circular groove on the tubing side of the flange. ISO-F flanges have holes for attaching the two flanges with bolts. The different ISO vacuum flange types are shown in Fig. 27.

The seals are rubber and are supported by an aluminium or stainless steel ring, which also serves to locate the seal on the flange. Various adaptor rings and components allow the whole range of LF components to be interchanged. ISO LF components are moderately priced, relatively easy to manufacture, but the clamping hardware is best bought commercially. The most common material used for the manufacture of the flanges and sealing joints is aluminium, although stainless steel is becoming more common.

A smaller type of ISO flange is the ISO Klein Flange (KF) also known as ISO quick flange (QF); see Fig. 27. These seals have a limited size range as they are designed specifically for the interconnection of small components, rather than larger diameter beamline type fittings. The typical method of sealing is a rubber seal, usually Viton, Buna, or silicone, between two flat sealing flanges with a tapered rim. The seal is supported in a retainer ring that is made from aluminium or stainless steel; however, plastic is sometimes used for insulated joints. The sealing surfaces are susceptible to scratching if placed on a rough surface. The flanges are clamped together using a specially designed clamp, which fits around

a) ISO-KF b) ISO-K c) ISO-F

FIG. 27. ISO vacuum flange types.

the tapered flanges squeezing them together as the clamp is finger tightened. The system parts are relatively inexpensive to purchase and manufacture.

11. BEAM TRANSPORT SYSTEM ELEMENTS

Beam transport systems include many electrostatic and magnet devices to focus, deflect, collimate, and change the shape of the ion beam. This section describes the commonly used elements of beam transport systems, and provides some recommendations for their effective use, and recommendations for diagnosing and rectifying common faults that may occur.

11.1 FOCUSING ELEMENTS

Electrostatic focusing lenses are used where the ion energy is low, typically a few tens of keV. They are typically found in ion sources and the low energy injection stage of accelerator systems. Their main purpose is to shape the characteristics of the ion beam to optimize its transmission through the accelerator. When the ion beam's energy is high, in the MeV range, the voltage required to produce a focusing action using electrostatic lenses becomes excessive so instead, magnetic lenses are used.

11.1.1 Einzel lens

The Einzel lens is the most used electrostatic focussing device. This lens consists of three coaxially aligned cylinders, with the voltage of the centre cylinder at a different polarity to the other two, as shown in Fig. 28. The polarity of the centre electrode depends on the polarity of the ion beam. The applied voltage does not change the beam energy because of the symmetrical potential of the electrodes.

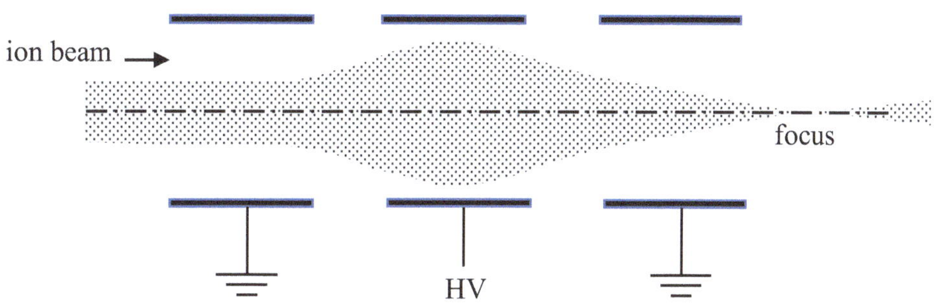

FIG. 28. Focussing action of an Einzel lens.

When used inside an ion source, the cylinder closest to the source of ions also serves as an extracting electrode. When an Einzel lens is used between the ion source and its pre-acceleration tube, the exit electrode of the Einzel lens is connected to the first electrode of the pre-acceleration tube to prevent any adverse shapes in the electrostatic fields that may produce unwanted transverse deflections. Einzel lenses can also be located near the first acceleration tube in the accelerator, to provide the best focusing condition for optimum transmission through the stripper canal. In this case, the last element of the Einzel lens is connected to one of the leading electrodes on the acceleration tube. As before, this also prevents any adverse transverse deflections from occurring.

11.1.2 Preventative maintenance of Einzel lenses

Einzel lenses are normally maintenance free devices, however some problems may occur if operated in adverse vacuum conditions. If the vacuum pressure is too high, the high voltages used can create a glow discharge between the electrodes, changing the shape and strength of the electric field lines causing a decrease in focussing action, and producing transverse beam displacements. If allowed to persist, this discharge can cause the smooth surface of the electrodes to be damaged, as well as depositing sputtered metal on the electrical insulators.

Contaminants in the vacuum such as water vapour or hydrocarbons can also facilitate the formation of electrical discharges, leading to surface contamination of the lens electrodes and insulators. Inspecting the condition of the focus electrodes is necessary after a long term of operation. A decolourization of the surface of the electrodes indicates the buildup of contaminants. Cleaning with organic solvents and polishing with fine abrasive paper can restore to a pristine surface. High voltage insulator feedthroughs need to also be inspected and cleaned to ensure reliable operation.

11.1.3 Magnetic quadrupole lens

A magnetic quadrupole lens is a lens with four magnetic poles as shown in Fig. 29.

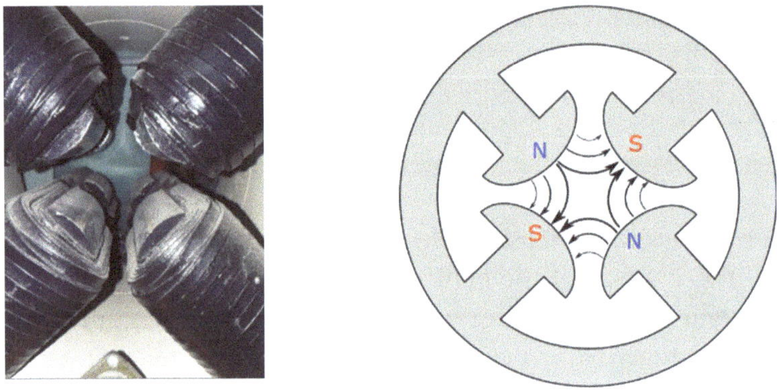

FIG. 29. Geometry of a magnetic quadrupole singlet lens.

A single quadrupole lens will only focus on one direction (either vertically or horizontally) so they are commonly used in pairs (doublets), with the magnetic poles of the second lens opposite to that of the first lens. This arrangement provides a net focusing of the ion beam in both vertical and horizontal directions. The strength of the magnetic fields, and the distance between the two lenses (drift section), determines the focal length of the quadrupole doublet.

The focusing action of a quadrupole doublet is illustrated in Fig. 30. Ions that travel along the central magnetic axis experience no deflecting force, there being no magnetic field at the central axis, but off-axis ions experience a deflecting force. In quadrupole 1, the magnetic forces will defocus the off-axis positive ions in the X-direction, and focus those in the Y-direction. As the beam travels through the drift space between the two quadrupoles, the diverging and converging trajectories of the ion beams continues. As the positive ion beam passes through quadrupole 2, the opposite magnetic polarity creates opposite focus/defocus forces; the X-direction is now focused, and the Y-direction defocused. The combined effect of the doublet is a net focusing in both directions.

Changing the relative electrical currents flowing through each of the lenses creates different focusing actions. Beams can be created with shapes such circular, elliptical, or line, with the major axis of these shapes being either vertical or horizontal. A quadrupole doublet lens is said to be balanced when the same electrical current is flowing through both lenses.

Care needs to be taken not to exceed the maximum current rating of the quadrupole's coils, as overheating may cause a short circuit in the coil windings. Cooling fans are typically used on high current quadrupoles.

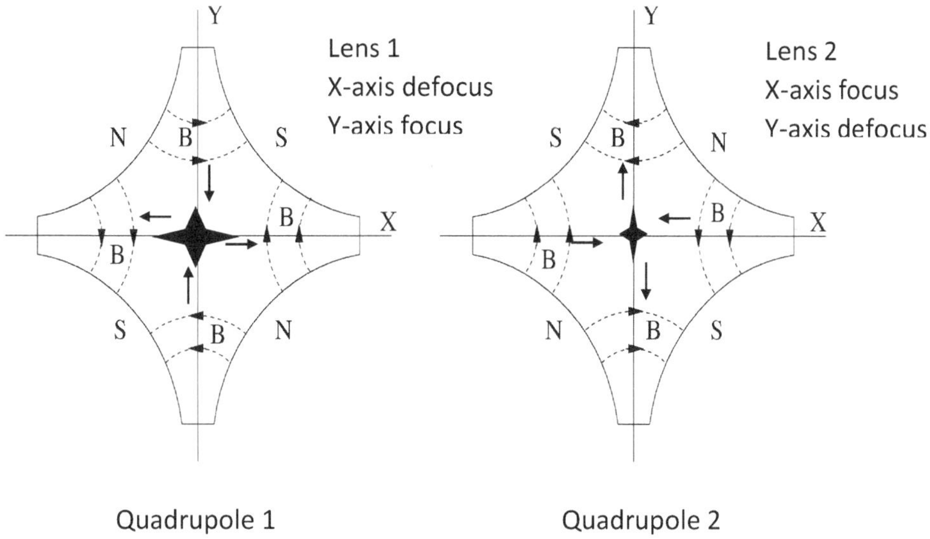

FIG. 30. Focusing action of a magnetic quadrupole doublet on positive ions.

11.1.4 Alignment of magnetic quadrupole lenses

A problem that is sometimes ignored with magnetic quadrupoles is the steering effect that can accompany the focussing actions. Steering is due to a misalignment of the central magnetic axis of the quadrupole lenses with respect to the central trajectory axis of the ion beam. Quadrupoles are typically set in place and aligned optically with the nominal beam axis. The effective magnetic (or electrostatic) axis of the quadrupole might not be the same as the geometric or optical axis. The offset between these two axes is the cause of the steering action that accompanies a focussing action. To rectify this misalignment, the quadrupole's orientation is incrementally changed until the steering effect decreases to acceptable levels.

To examine the alignment and focusing action of the quadrupole lenses:

— Insert a beam viewer or beam profile monitor after the lens system near its focal position.
— With the quadrupole lens inactive, pass an ion beam through its centre. The position of the beam on the beam viewing device is the reference position that needs to remain unchanged when the quadrupole is energized.
— Increase the current on one the quadrupole lenses while observing the beam shape. The beam shape will decrease in one direction and increase in the other. Repeat this exercise with the other lens of the doublet. A non-changing beam shape indicates that a problem exists with that lens.
— Next, while viewing the beam shape and position, incrementally increase the currents in both quadrupoles in unison. Ideally, the beam will become focussed without any accompanying steering action. If the beam moves away from the reference position, then a realignment of the quadrupoles may need to be done which is now described.
— Quadrupole magnets are mounted on a frame that usually has alignment jacks on each corner. Using quarter turns of the alignment jack bolts, tilt the quadrupole frame to correct for the steering angle. This may take many small adjustments and can become time consuming and frustrating, but it is much better to make numerous small changes rather than a few large changes. A well-aligned quadrupole will produce increased beam currents, less aberrations, and better reproducibility in experimental settings from day to day.

Defective focusing action of a quadrupole lens usually indicates a problem with the coil windings, or the connections. Coil windings are usually not repairable, so the whole quadrupole will need to be replaced.

11.1.5 Electrostatic quadrupole lens

Electrostatic quadrupoles are simpler in construction than their magnetic counterparts. They consist of four metal hyperbolic shaped poles, each pole being opposite in polarity to the adjacent pole. Figure 31 shows the geometrical arrangement of an electrostatic quadrupole. Their focusing action is not as strong as a magnetic quadrupole and needs high voltage power supplies with electrical feedthroughs penetrating the vacuum system. A high vacuum is required inside the lens to prevent corona discharges and other deleterious effects as previously mentioned for Einzel lenses.

Electrostatic quadrupoles are used in doublet or triplet arrangements, and have analogous focussing properties to magnetic quadrupoles.

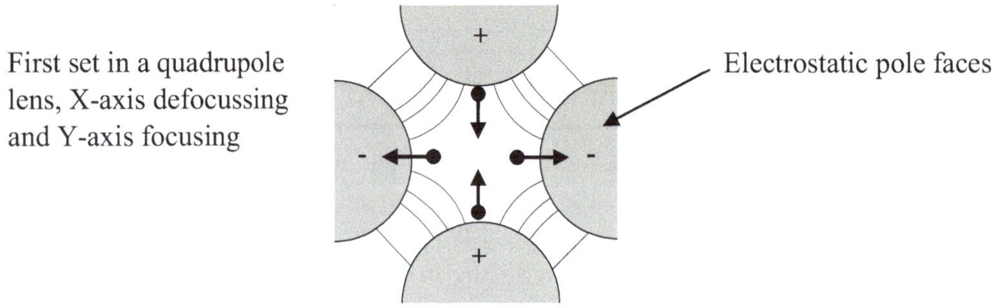

FIG. 31. Focussing action of an electrostatic quadrupole singlet lens on positive ions.

11.2 BEAM DEFLECTING ELEMENTS

Beam deflecting elements can be either electrostatic or magnetic. Electrostatic deflectors are typically used for low energy ion beams, and magnetic deflectors used for high energy beams.

11.2.1 Electrostatic deflectors

The simplest type of electrostatic deflectors consists of two sets of two parallel plates separated by a small gap of typically a few centimetres, as shown in Fig. 32. This device is located inside the vacuum system. High voltage feedthroughs into the vacuum system connect the plates to an external power

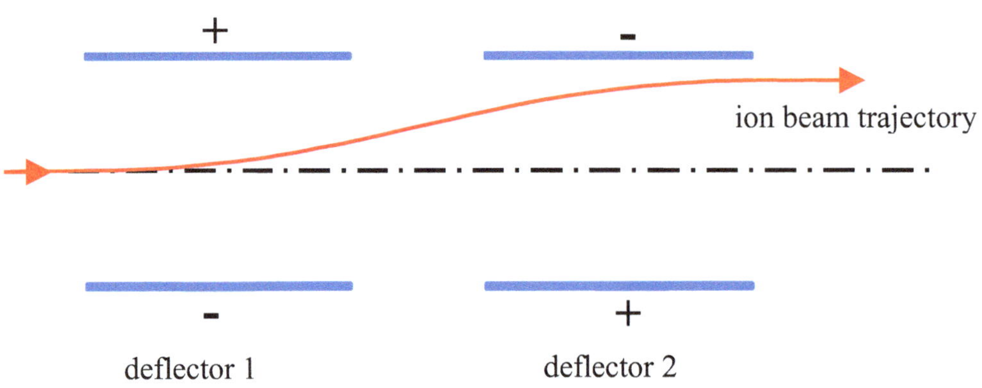

FIG. 32. Trajectory of a negative ion beam in an electrostatic steerer doublet.

supply. The electric field created between the deflector plates creates the deflecting force on the charged particles.

A common arrangement is to use a combined deflector with a second set of deflecting plates, identical to the first set, but with opposite polarity, and placed after the first set. This second deflector will return the deflecting angle back to 0°, producing an ion beam trajectory that is offset, but parallel to its initial trajectory.

11.2.2 Corrective maintenance of electrostatic deflectors

The most common problems that can occur with electrostatic deflectors are broken electrical wires inside of the deflector chamber, or the surface of electrical insulators being coated with a conducting layer. In both cases, the deflector needs to be removed for repair and cleaned.

11.2.3 Magnetic deflection

Magnetic deflectors are electromagnets, either in the form of a large analysing or switching dipole magnets. These magnets are made of two iron core poles with coil windings, separated by a nominal space (pole gap). In the pole gap is a vacuum chamber (magnet box), which has vacuum beam line inlet and outlet ports. The magnetic poles are supported by large sections of steel, which confine the effective magnetic field to the centre section of the magnet. Magnets can be designed to bend ion beams from a few degrees, to angles greater than 90°. Magnet vacuum boxes are often made of stainless steel, with welded ports that radiate outwards from an axis near the pole centres. Because the thickness of material used in the construction of the magnet boxes is relatively thin, distortions can occur during the welding process. The centre axis of the magnet box ports might not necessarily correspond to axis that the ion beam travels along. Any alignment of the beam transportation system needs to be based on the actual trajectory that the ion beam takes, rather than some assumed geometrical position of a mechanical component.

The magnetic flux inside the magnets is typically measured with a Hall probe. There is usually insufficient space between the pole faces and the magnet box to insert the Hall probe sensor, however its placement is not critical. Provided it is rigidly fixed in a reproducible position, the magnetic field sensor will read a value proportional to the actual field that is acting on the ion beam. A calibration table can be constructed of the measured Hall probe flux versus the mass-energy product of the ion beam.

11.2.4 Corrective maintenance magnetic deflectors

Large electromagnets require water cooling to keep their coil windings below their maximum operating temperature. Excessive temperatures can damage the insulation between the coil windings, creating short or open circuits. Water flow switches need to be installed in the water cooling lines and interlocked to the power supplies so that a low flow condition will disable the magnet power supply. Thermal sensors can be used as additional protection, to shut down the applied power if unacceptable operating temperatures are reached.

If the magnet power is on, and there is no magnetic field produced, then either there is no current being applied to the coils, or there is a defect in the coils. A no current condition usually arises from interlock problems, either because the interlock is active due to a fault condition, or the interlock itself has failed. A systematic check of the interlocks is required to isolate the fault condition. If a computer is used to control the magnetic power supply, then defective communication command signals are usually the fault.

Suspected problems with the magnet coil windings are usually determined by measuring the cold resistance of the windings and comparing it with the manufacturer's specifications. Sometimes, problems with coil windings only appear after the magnet has been operational for some time. This can arise from thermal expansion of parts of the coil windings creating changed electrical resistivities, or from a defect in the magnet power supply when its components become warm. In-situ monitoring of the voltage and current being applied to the magnet can help isolate the problem.

11.3 WIEN FILTER

A Wien filter, or E×B filter, is a velocity filter to select particles based on their speed. If all particles entering the Wien filter have the same energy, then selecting particles based on their speed is equivalent to selecting particles based on their mass, i.e. it is also a mass filter. The geometrical arrangement of a Wien filter is shown in Fig. 33.

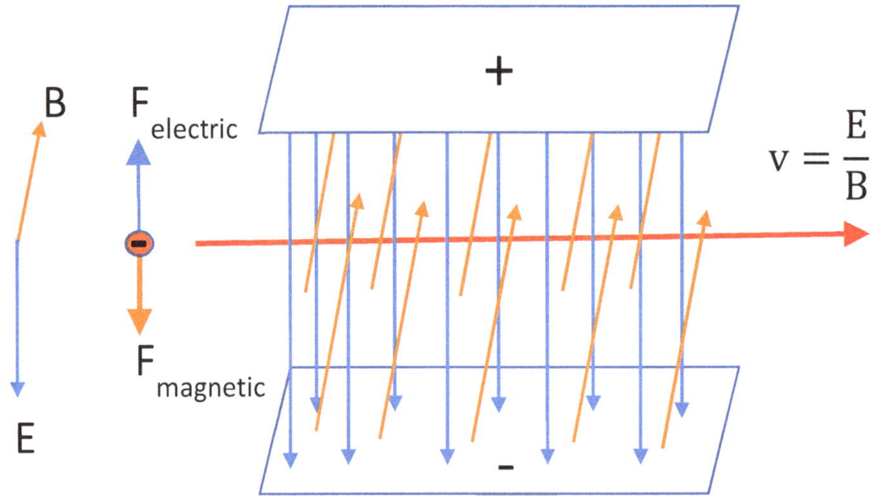

FIG. 33. Transmission of a negative ion through a Wien filter. The ion is undeflected when its speed v = E/B, where E is the electric field strength and B the magnetic field strength

The Wien filter consists of crossed electric and magnetic fields, with the field directions at 90° to each other. With this arrangement of the fields, the electric field will create a force $F_{electric}$ on the ion in one direction, and the magnetic field a force $F_{magnetic}$ in the opposite direction. When the speed of the ion is equal to the ratio E/B, where E is the electric field strength, and B is the magnetic field strength, both forces are equal and the ion pass through the filter undeflected. Adjusting the strengths of the magnetic and electric fields selects the velocity of the transmitted ion.

Maintenance of Wien filters includes careful voltage conditioning before use, regular cleaning of the interior high voltage surfaces, and operation only at high vacuum pressures. Conditioning of these devices needs to be done with the magnetic field on to ensure trapped gas molecules are drawn out of small voids in the interior mechanical structures. If the magnetic field is off during conditioning, a rise in vacuum will be seen when the high voltage is applied. This may cause electrical discharges and possible damage to the high voltage plates. All insulators need to be kept clean, and all terminals on the insulators need to have rounded edges to reduce the possibility of corona discharges occurring.

If the high voltage plates require cleaning, then fine abrasive paper, used dry, can be first used to remove damaged areas then followed by polishing with a metal polish. All residues from the polish need to be removed with a solvent and thoroughly dried before returning the device into service.

11.4 BEAM IMAGING DEVICES

Beam imagining devices are used to observe the location and shape of the ion beam. They can be either electrical readout devices (beam profile monitors) or optically viewed with the naked eye or CCTV. Caution need to be exerted when using naked eye viewing devices due to potentially high radiation doses that can be received.

11.4.1 Beam profile monitors

Beam profile monitors (BPMs) produce an electrical readout signal that is representative of the beam's shape, relative X and Y position, and intensity, which are displayed on an oscilloscope. These devices can monitor beams of electrons, ions, and energetic neutral particles. BPMs use either a rotating helical wire or a vibrating 'Y' wire to generate the readout signal.

The helical wire method uses a helical shaped wire mounted onto a rotary drive, as shown in Fig. 34. The rotating helical wire intersects and transverses the beam horizontally; then after 180° of rotation, it intersects and transverses the beam vertically. A cylindrical collector around the grounded wire collects the beam-induced electrons emitted from the wire, producing a signal proportional to the beam intensity profile in the X and Y directions.

BPM controllers accept inputs from multiple BPMs and display the selected BPM on an oscilloscope. They also provide a reference signal to indicate the beam axis centre (the fiducial marks) for each BPM. The signal generated from a BPM intersecting a well-defined symmetrical beam of uniform cross-sectional density will appear as two peaks of similar amplitude on the oscilloscope display, as shown in Fig. 34. If the centres of these peaks are not aligned with the fiducial marks, then the beam may be off axis. If the peaks are different heights, then the beam shape is not symmetrical in the X and Y directions.

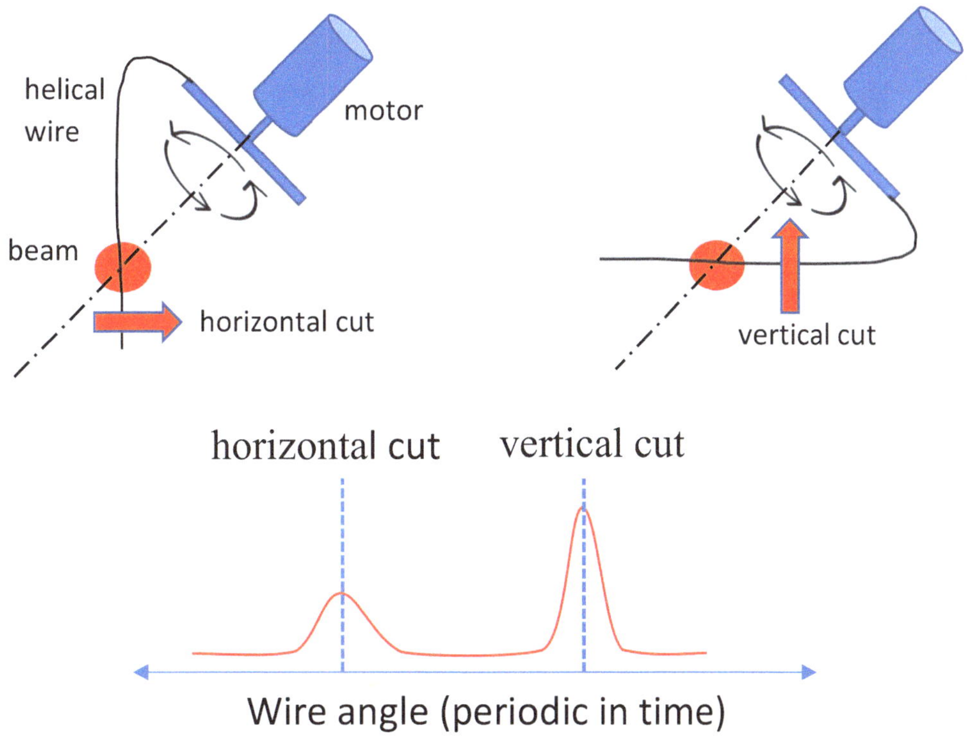

FIG. 34. Geometrical arrangement and electric readout of a helical wire BPM.

When the rotating drive is switched off, the motor shaft parks in a position where the helical wire is not in the path of the beam. A magnetically coupled drive eliminates rotating vacuum seals.

The vibrating 'Y' shaped wire method uses a 'Y' shaped wire that straddles the beam. In the switched off state, the wire is not in the path of the beam but when on, it vibrates back and forth intersecting the beam. It generates a signal in a similar way to the helical wire.

When BPMs are not being used, they need to be switched off to prolong the life of the motor bearings. BPMs have sensitivities as low as 1 nA. To maintain a high signal to noise ratio, it is important that the signal wires to the controller take paths which are free of any potential electrical interferences. This is especially important when working with very low beam currents.

11.4.2 Beam viewers

A beam viewer provides a visual display of the beam profile. A variety of beam viewers exist, but the most common type is the quartz viewer. In this type, an actuator or manually operated device is used to insert a piece of flat quartz, typically an inclined disk, into the path of the beam. The ion beam causes a blue fluorescence in the quartz providing a visual image of the beam profile. This image will always appear larger than the actual beam as there is some penumbra caused by light and electron scattering at the edges of the beam image. Care needs to be taken when using viewers as significant levels of radiation may be produced. Radiation levels need to be checked before visually observing the viewer. Video cameras have been successfully used in conjunction with the quartz viewers so that the beam can be viewed remotely.

The length of time quartz is exposed to the beam needs to be limited depending on the beam power. The quartz will eventually produce a red heat spot that forms at the core of the ion beam, which is the region of highest current density. A red spot indicates that the quartz is at a high temperature. It then need to be immediately removed from the beam to prevent damage to the quartz. With low power beams, it is sometimes necessary to allow a red spot to form, as the core of the beam is not always evident amongst the much larger diameter blue fluorescence.

Quartz is the preferred material in most situations due to its ability to withstand transient high heating power loads. Other fluorescing materials can also be used. CaF_2 crystal is a good fluorescent material for low beam currents. Alumina also fluoresces however as with many materials the intensity of the fluorescence decreases with use as the surface accumulates damage from ion beam irradiations.

11.4.3 Faraday cups

A Faraday cup is a device to quantitatively measure beam current. In its simplest form, it consists of a tantalum cylinder located inside the beamline, open at one end and closed at the other, as shown in Fig. 35.

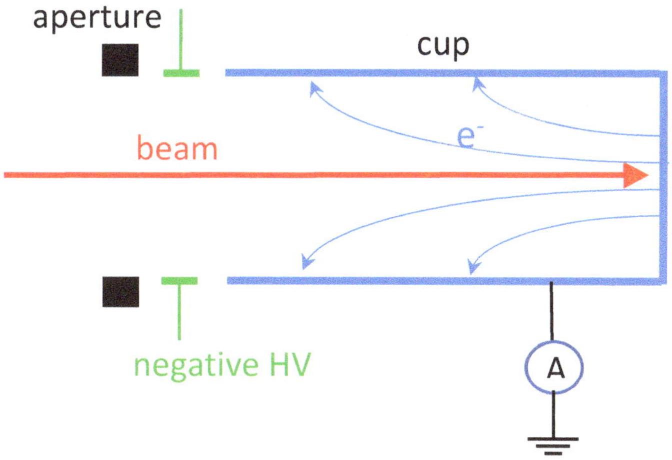

FIG. 35. Schematic of a Faraday cup with electron suppression.

When the beam enters the cylinder, the ions are collected, and the ion beam current is measured with a microammeter. The cylinder is typically made of tantalum which can withstand continuous heat loads of a few tens of watts without cooling, and which minimizes any neutron production when proton or helium beams are used. A pneumatic actuator inserts the cup into the beam for measurement and retracts it from the beam when not in use.

For accurate and reliable measurements, two potential sources of error need to be minimized: backscattered ions and secondary electrons leaving the cup. Both are current sources in addition to the ion beam current. Ions entering the cup can be backscattered 180° and exit the entrance aperture, resulting in an incomplete collection of all the incident ions. To minimize the proportion of backscattered ions escaping, long cups need to be used. It has been empirically found that the length of the cup needs to be at least 6–10 times the inside diameter. When the ion beam strikes an interior surface of the cup, low energy secondary electrons are emitted. If these secondary electrons can escape from the cup, the effect will be to introduce another current source other than the incident ion beam. To prevent electrons from escaping, long cups are used but more importantly, an electron suppression annulus is placed at the entrance of the cup and biased at a high negative voltage (~ -150 V) pushing the secondary electrons back into the cup.

An additional electron suppression feature of Faraday cups is an aperture or guard ring at the entrance to the cup, located in front of the electron suppressor. The internal diameter of this aperture ring is the same as the internal diameter of the cup, and its external diameter at least as large as the cup's outside diameter. The ring is used to define the area of beam entering the Faraday cup, and to prevent any fringe ions from striking the exterior surface of the cup or the suppressor creating an extraneous current source.

Wires from the cup need to be shielded to ensure they do not short circuit to ground, and to guard against accidental exposure to the ion beam to prevent the production of secondary electrons. In a vacuum system where the ion beam may intermittently strike the wiring, hollow cylindrical ceramic beads are the best shielding, as they are good electrical insulators, heat tolerant, and have low outgassing properties. In situations where the beam cannot strike the wiring, then Teflon shielding is a suitable and preferable alternative.

11.4.4 Beam defining slits and apertures

Beam slits are used for defining the beam to a particular size or confining its trajectory to a particular direction. They are adjustable from outside of the beam line, and usually employ some form of micrometer as a positional indicator. Beam defining slits are generally not intended to intercept large amounts of beam, so their power dissipation rating is low. Apertures and slits need to be manufactured from materials such as carbon or tantalum which do not produce high levels of radiation from nuclear reactions with the ion beam. A radiation survey is needed to check for potentially high radiation levels around the slits when a new ion beam or experimental condition is being used for the first time. If the slits are electrically isolated, as would be the case when the ion beam current on the slits is to be monitored, they are not to be disconnected from an electrical discharge path. Unconnected slits will accumulate electrical charge from the intercepted ion beam creating an electrostatic field that can deflect the beam. Some apertures are water cooled to remove the heat deposited by the intersecting beam. A radiation survey is necessary when removing apertures and slits as they may become activated after prolonged use.

12. COMPUTER CONTROL

In the early days of electrostatic accelerator development, the control of the accelerator system and the associated equipment was accomplished through a console containing a large number of switches, knobs, gauges, dials and indicator lights. The console was situated relatively far from the machine, requiring long cables making the operation susceptible to interference and ground loops. Normally, each element of the accelerator system was controlled by a custom chassis in the console. Since there may be hundreds of control elements, the cost of the machine control represented a significant portion of the total cost of an electrostatic accelerator. High voltage discharges may create high current spikes that can burn electronic circuits from adjacent chasses. The biggest disadvantage of the manual control of the accelerator was that the startup and the shutdown procedure required extremely skilled operators taking several hours dedicated to machine starting.

Computer control systems have alleviated much of the problems of a knob-based control system. The computer control system is normally provided by the manufacturer with the machine. Custom implementations are common, since they allow an easier upscaling and adaptation of the accelerator system to the needs of different experimental set-ups.

12.1 CONTROL ELEMENTS AND SIGNALS

Generically, there are two basic types of signal for control elements, analogue and digital, and two directions for the signals, read and write (set).

Analogue signals are required to control elements that can have continuous values between a minimum and maximum value. For example, in a knob-based system, the position of a knob establishes a voltage output from a resistive divider that is proportional to the knob position. In a computer control system, a graphical element from a graphical user interface is determining the output voltage value of a digital to analogue converter that is proportional to the user set value. Readout signals are simply gauges in a knob-based system, while in computer control systems they are labels or graphical gauges that display the digitally converted value of an analogue voltage from an analogue to digital converter.

Digital signals have only two values, ON or OFF, and are an equivalent to switches and lights in a knob-based system. In computer control systems, buttons, and labels (or coloured controls) are used. To exemplify, the list below covers the most used analogue and digital elements.

Analogue signals:
- Power supply set and read values (current or voltage): filament currents, extraction voltages, electrostatic element voltages, magnet currents, terminal voltage;
- Pressure gauges, temperature sensors, magnetic field sensors;
- Mechanic actuators: target position in a carousel, leak valves position (i.e. stripper);
- Dedicated electronics signals (e.g. GVM);
- Radiation level readouts.

Digital signals:
- Vacuum valve switches:
- Flow detection for cooling (ion sources, magnets or electronics);
- Pump switches:
- Threshold indicators (good or bad vacuum for example);
- Current direction in switching magnets (positive or negative);

— Door positions (opened or closed);
— Radiation thresholds and alarms.

Each of these elements can have a set and a readout, or only one of them, depending on the type of element that is controlled. For example, a pressure gauge can have only a read value, since it can only read the pressure. A current supply has typically a set and a readout value. An electrical switch normally has only a set value.

The number of elements to be controlled in an electrostatic accelerator may be several hundred, depending on the number of ion sources, the level of control of the accelerator, and the instrumentation required by experimental stations.

12.2 HARDWARE

A computer control system requires specialized hardware and software in order to accomplish its goal. There is no specific hardware dedicated to electrostatic accelerators in particular, it is rather a choice of the manufacturer or the facility for specific existing commercial hardware, with possible addition of in-house developments.

The most direct hardware implementation is to set up enough analogue and digital input and output computer cards, and to adapt their signals to the one required by the hardware on the accelerator side. Such implementation would still have some of the problems of the old knob-based system, with possible electromagnetic interference, ground loops, and current spikes. To overcome these problems, several scenarios can be implemented:

— An optical isolation for each signal. This requires conversion modules on each side of the optical cable. For most of the analogue signals, a frequency is transmitted in order to avoid attenuation issues related with the fibre lengths. It is a rather costly solution but commonly encountered.
— Using serial standard communication protocols (RS232, RS485) with converters (analogue and digital) on the device side.
— Signal multiplexing with an optional optical separation only of the multiplexed part (using controller area network, Ethernet or other protocols).
— A distributed control system, having more than one control computer, communicating via a software bus. Scenarios.

Usually, a combination of the above is implemented. For example, the analogue elements can be controlled by cards inside a main control computer, but the digital elements can be controlled by a programmable logic controller that communicates with the main computer via a serial protocol.

12.3 SOFTWARE

The first software component common to any control system software is that which is used to interface the hardware units with the operating system (device drivers). As a good practice, device drivers need to be stored safely as, if the computer needs to be changed or replaced, those drivers are needed. In case of updating or changing the operating system, be sure to have up to date device drivers.

The other software components required by the control system can be generically divided in three parts:

— The graphical user interface, which is the graphical part that the operator uses to set and control any element;
— A logic part that contains automatic procedures or calculation required by the operation;
— A log part that is used to store and display operational values.

Normally, the accelerator manufacturers provide their own in-house software to cover those parts. At the facility level, software packages such as EPICS, LabVIEW or Tango can be used to develop custom control systems.

Some common properties of the three software components above are now described, and it may serve as a guide for the operator to understand the functionality of the control software used in their accelerator facility, or to help a software developer to implement or expand their computer control system.

The first part, the graphical user interface, is a particularly important part, since it is what the operator is seeing. A good accelerator computer control system has the minimum possible number of windows for the operator to interact with. A single display monitor is typically sufficient for operation. The windows, or the tabs with graphical elements, are typically divided in operational groups as follows:

— Ion sources;
— Low energy injection;
— Accelerator;
— High energy focusing and energy selection;
— Experimental station beam transport.

Separate windows can be used to monitor the vacuum system, the beam current at different measurement points along the beamline and other locations. Error conditions can be signalled with the label background in different colours, normally green indicates a good state, red for error states.

The response time of an accelerator control system is an important parameter to consider. In practice, the readout frequency of any displayed control needs to be faster than four times per second. A lower frequency makes beam tuning difficult.

Control of items involving safety or radiation protection need to be independent of the operating computer control system. Being safety related, they need to be operational regardless of the state of the computer control system and in general, not able to be overridden or disabled by the computer control system. It is reasonable to monitor such systems by the main control computer, and to include restricted overrides based on higher levels of password authentication only by authorized personnel. Such items include radiation interlocks, vacuum interlocks, and high-voltage interlocks. Primary control of these items needs to be hardwired systems or controlled by industrial grade programmable logic controllers with default values assuring safety.

The second part is the logic that contains automatic procedures or calculations required for the operation. It contains:

— Automatic ion source starting and shutdown procedure;
— Magnet current calculation for mass/charge selection of magnetic devices;
— Conditioning of the accelerator tube;
— Terminal voltage calculation based on the required energy, mass, charge, and calibration values;

— Parametric scans, for example displaying the current in a Faraday cup while scanning the current in the low energy magnet;
— Filament outgassing procedure.

The logging part of the control software is important for the efficient setting up and restoring the system when needed. Considering that several hundreds of parameters are stored, several times per second, the log files can be rather large. Storing the data in a binary format is a way to minimize the storage space for these log files. The control software needs to have the necessary utilities for later parameter visualizations using data from the log files.

If a computer control system is to be upgraded or changed, the following factors need to be taken into consideration:

— The stability of the control system: it is convenient to test smaller scale models or partial implementations to check for correct functioning. Crashing software can be a serious problem during everyday operation.
— Support for hardware: check for the availability of and support for device drivers.
— Long term support for the solution: basically, choosing a configuration with a large developer community with a development plan for as many years as possible.
— Estimated operator learning curve for the new control system.

REFERENCES

[1] INTERNATIONAL ATOMIC ENERGY AGENCY, Accelerator Knowledge Portal (2023), https://nucleus.iaea.org/sites/accelerators/Pages/default.aspx

[2] THOMAS JEFFERSON NATIONAL ACCELERATOR FACILITY, Standard Operating Procedure (SOP), https://www.jlab.org/accel/inj_group/safety/PROCEDURES/SOP_ITS_South_060729.pdf

[3] INTERNATIONAL ATOMIC ENERGY AGENCY, Leadership and Management for Safety, IAEA Safety Standards Series No. GSR Part 2, IAEA, Vienna (2016), https://doi.org/10.61092/iaea.cq1k-j5z3

[4] INTERNATIONAL ATOMIC ENERGY AGENCY, Occupational Radiation Protection, IAEA Safety Standards Series No. GSG-7, IAEA, Vienna (2018).

[5] INTERNATIONAL ATOMIC ENERGY AGENCY, Radiation Safety in the Use of Radiation Sources in Research and Education, IAEA Safety Standards Series No. SSG-87, IAEA, Vienna (2024), https://doi.org/10.61092/iaea.0f4a-dm2n

[6] EUROPEAN COMMISSION, FOOD AND AGRICULTURE ORGANIZATION OF THE UNITED NATIONS, INTERNATIONAL ATOMIC ENERGY AGENCY, INTERNATIONAL LABOUR ORGANIZATION, OECD NUCLEAR ENERGY AGENCY, PAN AMERICAN HEALTH ORGANIZATION, UNITED NATIONS ENVIRONMENT PROGRAMME, WORLD HEALTH ORGANIZATION, Radiation Protection and Safety of Radiation Sources: International Basic Safety Standards, IAEA Safety Standards Series No. GSR Part 3, IAEA, Vienna (2014), https://doi.org/10.61092/iaea.u2pu-60vm

[7] INTERNATIONAL ATOMIC ENERGY AGENCY, Radiation Protection in the Design of Radiotherapy Facilities, Safety Reports Series No. 47, IAEA, Vienna (2006).

[8] SPELLMAN HIGH VOLTAGE, Spellman High Voltage Reference Manual, https://www.spellmanhv.com/-/media/en/Technical-Resources/Manuals/hv-ref-manual.pdf

[9] SOLVAY, Sulphur Hexafluoride, https://www.solvay.com/sites/g/files/srpend221/files/tridion/documents/SF6-Sulphur-Hexafluoride-EN.pdf

[10] UNITED STATES ENVIRONMENTAL PROTECTION AGENCY, Sulfur Hexafluoride (SF6) Basics, https://www.epa.gov/eps-partnership/sulfur-hexafluoride-sf6-basics

[11] UNITED STATES ENVIRONMENTAL PROTECTION AGENCY, Byproducts of Sulphur Hexafluoride (SF6) Use in the Electric Power Industry, https://www.epa.gov/sites/production/files/2016-02/documents/sf6_byproducts.pdf

[12] UNITED STATES DEPARTMENT OF LABOR, Occupational Safety and Health Administration, https://www.osha.gov/laws-regs/regulations/standardnumber/1910/1910.134

[13] NORTON, G.A., "Box 10: Earthquake Protection–for Pelletrons", Electrostatic Accelerators: Fundamentals and Applications (HELLBORG, R., Ed.), Springer-Verlag, Berlin and Heidelberg (2005) 372-373

[14] RUĐER BOŠKOVIĆ INSTITUTE, RBI accelerator facility and the earthquake (and the corona), https://www.irb.hr/eng/Divisions/Division-of-Experimental-Physics/Laboratory-for-ion-beam-interactions/Articles/RBI-accelerator-and-earthquake

[15] TOMMASINI, D., "Dielectric insulation and high voltage issues", (Proc. CERN Accelerator School - CAS 2009: Specialised Course on Magnets, Bruges, Belgium, 2009), CERN Document Server, https://cds.cern.ch/record/1342701/files/335.pdf

[16] LEUNG, K.N., BACHMAN, D.A., MCDONALD, D.S., YOUNG, A.T., "Ion sources for cyclotron applications", (Proc. 13th Intern. Conf. Cyclotrons and their Applications, Vancouver, Canada, 1992), World Scientific, Singapore (1993), 312-315 and CERN Document Server, https://accelconf.web.cern.ch/c92/papers/viii-03.pdf

[17] DONNALY, B.L., Process for producing negative helium ions, US Patent 3374384, Priority May 5, 1966, Patented Mar. 19, 1968; https://patentimages.storage.googleapis.com/5a/1c/dc/aabcb4ef8093c4/US3374384.pdf

[18] SCHLACHTER, A.S., Formation of negative ions by charge transfer: He^- to Cl^-, AIP Conference Proceedings **111**, 300 (1984)

[19] VAN DE GRAAFF, R.J., Inclined-Field High-Voltage Vacuum Tubes, US Patent 3308323, Priority May 25, 1961, Patented Mar. 7, 1967; https://patentimages.storage.googleapis.com/14/19/86/2258e0d148d78c/US3308323.pdf

[20] BAGLIN, V., BOJKO, J., GRÖBNER, O., HENRIST, B., HILLERET, N., SCHEUERLEIN, C., TABORELLI, M., "The Secondary Electron Yield of Technical Materials and its Variation with Surface Treatments", (Proc. 7th European Particle Accelerator Conf., Vienna, Austria, 2000), CERN Document Server, https://cds.cern.ch/record/466534/files/lhc-project-report-433.pdf

[21] HIGH VOLTAGE ENGINEERING EUROPA B.V., Duoplasmatron Ion Source, https://www.highvolteng.com/media/Leaflets/model_358__duoplasmatron_ion_source.pdf

[22] DOW CHEMICAL COMPANY, Syltherm XLT Heat Transfer Fluid Technical Manual, https://www.dow.com/en-us/document-viewer.html?ramdomVar=7840425133539237l3&docPath=/content/dam/dcc/documents/en-us/app-tech-guide/176/176-01434-01-syltherm-xlt-heat-transfer-fluid-technical-manual.pdf

[23] HARVEY, A.H., HRUBÝ, J., MEIER K., Improved and Always Improving: Reference Formulations for Thermophysical Properties of Water, J. Phys. Chem. Ref. Data **52**, (2023) 011501, https://doi.org/10.1063/5.0125524

[24] LIGHT, T.S., LICHT, S., BEVILACQUA, A.C., MORASHC, K.R., The Fundamental Conductivity and Resistivity of Water, Electrochemical and Solid-State Letters **8** (2005) E16, https://iopscience.iop.org/article/10.1149/1.1836121

[25] CHIANG, H-W.D., KUAN, C-P., LI, H-L., Turbomolecular pump rotor-bearing system analysis and testing, Journal of Vacuum Science & Technology A **27**, (2009) 1196, https://doi.org/10.1116/1.3179157

LIST OF ABBREVIATIONS

AC	alternating current
BPM	beam profile monitor
CCTV	closed circuit television
CF	conFlat
CPO	capacitive pick-off
DC	direct current
GVM	generating voltmeter
ISO	International Organization for Standardization
LN2	liquid nitrogen
MSDS	material safety data sheet
OHS	occupational health and safety
PVA	polyvinylacetate
RF	radiofrequency
RGA	residual gas analyser
SOP	standard operating procedures

CONTRIBUTORS TO DRAFTING AND REVIEW

Bartha, L.	Institute of Nuclear Research of the Hungarian Academy of Sciences, Hungary
Charisopoulos, S.	International Atomic Energy Agency
Dytlewski, N.	Independent consultant, Australia
Fallon, J.	Australian Nuclear Science and Technology Organization, Australia
Fazinić, S.	Ruđer Bošković Institute, Croatia
Garton, D.	Australian Nuclear Science and Technology Organization, Australia
Joco, V.	Autonomous University of Madrid, Spain
Marinescu, L.	National Institute of Physics and Engineering, Romania
Skukan, N.	International Atomic Energy Agency
Vavpetič, P.	Jožef Stefan Institute, Slovenia
Westerfeldt, C.	Duke University, USA

Consultancy Meetings

Vienna, Austria: 18–20 December 2017

IAEA
International Atomic Energy Agency

CONTACT IAEA PUBLISHING

Feedback on IAEA publications may be given via the on-line form available at:
www.iaea.org/publications/feedback

This form may also be used to report safety issues or environmental queries concerning IAEA publications.

Alternatively, contact IAEA Publishing:

Publishing Section
International Atomic Energy Agency
Vienna International Centre, PO Box 100, 1400 Vienna, Austria
Telephone: +43 1 2600 22529 or 22530
Email: sales.publications@iaea.org
www.iaea.org/publications

Priced and unpriced IAEA publications may be ordered directly from the IAEA.

ORDERING LOCALLY

Priced IAEA publications may be purchased from regional distributors and from major local booksellers.

25-00819E

FSC
www.fsc.org
MIX
Papier | Fördert
gute Waldnutzung
FSC® C083411

Zeitfracht Medien GmbH
Ferdinand-Jühlke-Straße 7
99095 Erfurt, Deutschland
produktsicherheit@kolibri360.de